WITHDRAWN

HOW TO
USE
STATISTICS

HOW TO USE STATISTICS

STEVE LAKIN

ALWAYS LEARNING

PEARSON

Harlow, England • London • New York • Boston • San Francisco • Toronto • Sydney
Auckland • Singapore • Hong Kong • Tokyo • Seoul • Taipei • New Delhi
Cape Town • Sao Paulo • Mexico City • Madrid • Amsterdam • Munich • Paris • Milan

Pearson Education Limited
Edinburgh Gate
Harlow
Essex CM20 2JE
England

and Associated Companies throughout the world

Visit us on the World Wide Web at:
www.pearsoned.co.uk

First published 2011

© Pearson Education Limited 2011

ISBN: 978-0-273-74387-3

British Library Cataloguing-in-Publication Data
A catalogue record for this book is available from the British Library

Library of Congress Cataloging-in-Publication Data
Lakin, Steve.
  How to use statistics / Steve Lakin.
    p. cm.
  ISBN 978-0-273-74387-3 (pbk.)
  1. Mathematical statistics--Textbooks. I. Title

  QA276.12.L35 2011
  519.5--dc22

                                    2011004370

10 9 8 7 6 5 4 3 2 1
15 14 13 12 11

Typeset in 10/13pt Din Regular by 3
Printed by Ashford Colour Press Ltd., Gosport

Smarter Study Skills

Instant answers to your most pressing university skills problems and queries

Are there any secrets to successful study?

The simple answer is 'yes' – there are some essential skills, tips and techniques that can help you to improve your performance and success in all areas of your university studies.

These handy, easy-to-use guides to the most common areas where most students need help, provide accessible, straightforward practical tips and instant solutions that provide you with the tools and techniques that will enable you to improve your performance and get better results – and better grades!

Each book in the series allows you to assess and address a particular set of skills and strategies, in crucial areas of your studies. Each book then delivers practical, no-nonsense tips, techniques and strategies that will enable you to significantly improve your abilities and performance in time to make a difference.

The books in the series are:

- *How to Write Essays & Assignments*
- *How to Write Dissertations & Project Reports*
- *How to Argue*
- *How to Improve your Maths Skills*
- *How to Use Statistics*
- *How to Succeed in Exams & Assessments*

For a complete handbook covering all of these study skills and more:

- *The Smarter Study Skills Companion*

Get smart, get a head start!

Contents

Probability

Probability distributions

Hypothesis testing

Statistical tables

Summary, glossary and appendices

Preface and acknowledgements

Statistics – isn't that really hard and complicated maths? I can't do that!

If that's your first thought, don't worry; you're not alone.

However, you actually see and use statistics quite a lot, and it doesn't need to be something scary.

As a student, no matter what course you are doing, you will probably need statistical analysis. The aim of this book is to gently introduce you to the basic topics needed without blinding you with mathematical formulae, to take away any fears of the subject you may have, and to provide you with the basic skills in data, presentation, probability and statistics that you may well encounter during your studies.

Instead of a textbook that throws complex formulae at you and says 'learn this', I'm going to try to explain how and why we make the calculations we do. As you learn more, you might find that you can use statistical software packages to solve these problems, but it's important that you know what the buttons and commands in your software package actually *do* – not just know that you can press them, but understand what is really going on. This is really the aim.

I hope the book will build your confidence, and show you that this sort of statistics is relevant and useful to your studies and your life.

This book would not have been possible without the support of my former colleagues at the University of Glamorgan, and my current colleagues at the University of Greenwich, to whom I am very grateful. I should also like to express my sincere thanks to my former PhD supervisor, Rick Thomas, at the University of Leicester, whose encouragement, inspiration and support will never be forgotten.

I should also like to thank Pearson Education for supporting this publication, and in particular Katy Robinson.

Finally, I should like to thank my family, whose constant support has always been there; special thanks to my wife Dasha for absolutely everything.

Steve Lakin

How to use this book

How to Use Statistics has been organised and designed to be as user-friendly as possible. Each chapter is self-contained and deals with a particular aspect of its subject matter, so that you can read the book through from end-to-end, or in sections, or dip into specific chapters as and when you think you need them.

At the start of each chapter you'll find a brief paragraph and a **Key topics** list, which lets you know what's included. There is also a list of **Key terms** at this point, and, should you be uncertain about the meaning of any of these, you will find definitions in the **Glossary** at the end of the book.

In this book we cover the basic ideas in data, measures, probability, testing and statistics, and also some other topics that are likely to be useful for you. It's not meant to be a comprehensive guide to the whole of statistics, other books do that; this is just something to give you the basics and make you see the relevance of the subject to whatever you are doing. You don't need any prior knowledge of statistics, and the tone is intended to be supportive and introductory.

Each chapter talks you through a particular topic, with examples illustrating the idea. Included in each chapter are a few 'smart tips' that we hope you will take on board and use to relate the topics to your studies and your daily life. Always try to keep in mind that statistics is meant to be useful and relevant!

At the end of each chapter is a set of exercises for you to try. Don't despair if you find some of them hard; the intention is to provide a mix of questions for each topic. For some of the more mathematical questions, examples (with solutions) are provided, but for the longer questions, they generally follow the pattern of those given in the chapter.

You can find solutions at the back of the book, but these are only the final answers. There is a companion website to this book, which you can link to from **www.smarterstudyskills.com**.

On the website you can find worked solutions to all the exercises, together with further questions (including some related to particular fields of study), contact details, and more. If you really don't understand a particular answer, go to the website to see the full worked solution, and see whether that helps you.

The book is meant to help you. Statistics isn't natural to a lot of people, but it's my hope that this book will take away some of its mystery, and make you see its relevance to your daily life, and to your studies. You *can* succeed with the subject, and I hope this book will help you do just that.

Smart tip boxes emphasise key advice to ensure you adopt a successful approach.

Information boxes provide additional information, such as useful definitions or examples.

 There are **Exercises** for you to do by yourself or in a group, which test whether you have understood the chapter, and give you an opportunity to put what you have learnt into practice.

 At the end of each chapter, a short **Summary** reinforces your learning and provides a reminder of the skills and understanding that you will have developed.

A note on mathematics, rounding, calculators and computer software

To be able to deal with statistics, you need a basic capability with mathematics. Although this book is self-contained, and tries to explain things with as little formal maths as possible, it does assume a basic level of maths knowledge in places.

If you are really maths-terrified, then I recommend you get up to speed first, before ploughing head-first into this book.

If you're really nervous about your maths, the companion title in this series *How to Improve Your Maths Skills* would be well worth reading first – it is intended to build up your confidence in the sort of basic mathematics you will encounter.

Also, once statistics gets complicated, computer software packages can really help; because the calculations get so long, mistakes are almost inevitable when done by hand. This book is not the place to tell you how to use particular software; there are dedicated books that can do that far better than I can in this book, which is meant to introduce the ideas of statistics. I shall give you a few simple Excel commands as an appendix, but if you need to use a computer package to deal with the statistics you require, then refer to a specialised textbook or guide.

When it comes to using a calculator, it's vital you know how to use your own calculator. Many calculators have their own statistical modes built in, and if you can work out how to use these, then great, but you must be able to at least perform complex calculations correctly. If you aren't sure about some aspects of your calculator, take the calculator test in the appendices. This contains calculations similar to examples in the book, and you need to be able to perform all of these correctly on your calculator. If you can't get the answers, then ask for help, consult the manual, get another easier-to-use calculator, use a computer, or do something else – just make sure you can get these answers!

Finally, as an important note, please don't round off your answers to some number of decimal places until the end. If you round off early in a calculation, you are introducing errors, which can grow and grow through the calculation, and you end up with your final answer being hopelessly inaccurate. Even as a simple example, if you round off $\frac{1}{3}$ as 0.3, then if you do $\frac{1}{3} + \frac{1}{3} + \frac{1}{3}$ you get $0.3 + 0.3 + 0.3 = 0.9$, rather than the actual answer of 1. This is a long way out – basically 10%. When doing multiple calculations, even small errors can be magnified. So please use fractions where possible, and only round at the end if you really have to.

INTRODUCTORY STATISTICS

1 | Introduction to statistics and data

An introduction to the basic concepts of statistics and data

Statistics is all about the interpretation of data. To begin, I shall first give a brief, informal overview of the major types of statistics, and discuss the different types of data that we may encounter.

Key topics
- Statistics
- Data

Key terms
statistics   data   data collection   descriptive statistics
inferential statistics   discrete data   continuous data
frequency distribution

● What is statistics?

Statistics is the branch of mathematics that deals with data. Data (technically a plural word; the singular is 'datum') is a collection of values. For most of what we do, it will be numerical data (such as the inflation rate, the number of bees in a colony, or the marks in a class test), but it can also take other forms (such as the political party a voter intends to vote for, the football team they support, and so on).

A collection of data is often referred to as a *data set* or *set of data*, but other words such as a *list* or simply *collection* are also often used. Don't worry too much about the words, just understand that we are referring to a collection of values.

Examples of data sets are:

- marks in a class test: 9, 2, 5, 8, 10, 3, 5, 8, 8, 9
- inflation rate: 2.1, 3.2, 4.1, 2.3, 5.1, 2.2, 0.5
- voting intention in a referendum: Yes, No, No, Yes, Yes, No

and so on. You will see many data sets as we go through the book.

There are three real branches of statistics: *data collection*, *descriptive statistics* and *inferential statistics*. Let us look at these concepts in a little more detail.

● Data collection

Data collection is all about how the actual data is collected. For the most part, this needn't concern us too much in terms of the mathematics (we just work with what we are given), but there are significant issues to consider when actually collecting data.

For data such as marks in a class test, this is fairly straightforward. Each student has a defined mark associated with them, so the marks are simply collected together to make the data set.

Sometimes, data is harder to collect. Counting the number of bees in a colony isn't easy, because they move and fly around; you may have to approximate in such cases.

Also, if you are collecting data, you need to be careful where you get it from. For example, suppose you want to conduct a poll on who people plan to vote for in an election. You can't realistically ask everyone in the whole country (the *population*), so you have to choose a representative *sample* of people. This isn't as easy as it sounds. In the mid 20th century, for example, polls were sometimes carried out by randomly calling people in the telephone directory. This sounds representative, but in those days only the richer people had telephones, and so you were asking only a particular section of society, who might well be more inclined to vote for one party rather than other. The same issue may apply with doing a poll by email today.

So there are issues in the collection of the data; you need to make sure that the data has been collected fairly before you go on to deal with it, and try to present it and make conclusions.

The words *population* and *sample* are used in general in statistics. The *population* is the entire set of data, and a *sample* is a (hopefully representative) subset of the population – so just 'some of' the data values. Why would we need a sample? As indicated above, it's usually unrealistic to get data from the entire population (it would be incredibly expensive and time-consuming, and also the population may be changing as you collect the data), so often we need to simply take a sample.

● Descriptive statistics

Descriptive statistics is the part of statistics that deals with presenting the data we have. This can take two basic forms – presenting aspects of the data either visually (via graphs, charts, etc.) or numerically (via averages and so on).

Common visual techniques that we shall discuss in Chapter 2 include graphs, bar charts, pie charts and more, but we shall focus mainly on numerical techniques such as averages and spreads.

The basic aim of descriptive statistics is to 'present the data' in an understandable way. If you simply write down every piece of data, it means little to someone who sees it; it needs to be summarised. Imagine if, on the TV news, they listed on the screen the votes of every single person interviewed by a polling company; it would just be a huge list of parties, and you couldn't arrive at any meaningful conclusion. Instead, you are presented with visual charts (a bar chart, say) to give, perhaps, the percentage of the vote each party has. In the 2010 General Election almost 30 million people voted. If each vote was simply written down and displayed, one after the other, you'd be totally lost; what happens is that a summary of votes is presented (for example as percentages: Conservative 36%, Labour 29%, Liberal Democrat 23%, Others 12%). This is an example of descriptive statistics – 'describing' or 'summarising' the overall data for people to understand.

● Inferential statistics

Inferential statistics is the aspect that deals with making conclusions about the data. This is quite a wide area; essentially you are asking 'What is this data telling us, and what should we do?'

For example, a council might be considering altering the speed limit on a main road, after a number of accidents. They might do this by surveying the speeds of cars (data collection) and then arrive at a conclusion as to whether the speed limit needs to be lowered (if, for example, a number of cars are driving too fast). Note, though, that this may not be the case; everyone might be driving at a perfectly acceptable speed, and the accidents are down to something other than speed (a blind spot or a pothole, for example). This is inferential statistics: take the data you have and make an 'inference' or 'conclusion' from it. We shall see much more of this later when we discuss things such as *hypothesis testing*, where we test to see whether the data supports a belief that we have.

● Discrete and continuous data

Data comes in two distinct types. *Discrete* data can take distinct values, which can be clearly identified and separated. An example of this is the score obtained by rolling a die, which can only take values of 1, 2, 3, 4, 5 or 6, with nothing in between, and all the scores can be distinguished. By contrast, *continuous* data can take any value. For example, when you measure the speed of a car, it could take any value, depending on how accurately you measure it – for example 31.2 or 48.28, or 48.281 – basically any value.

A good illustration is to consider the whole numbers (1, 2, 3, etc.), which are clearly distinct from each other (and so discrete), and the positive real numbers (every number you can think of, including decimals), which are continuous. If you draw a line of the whole numbers between 1 and 10 it looks something like

1 – 2 – 3 – 4 – 5 – 6 – 7 – 8 – 9 – 10

and it is clear that the numbers are distinct. But if you draw a line of the real numbers between 1 and 10, it's just a single straight line – instead of taking one of the 10 discrete values as above, you can now take any value in the range, for example 1.75, 2.895, 9.238984 and so on.

———————————————————————

1   10

Continuous data will almost always be approximated (to 1 decimal place say, as in the speed of a car, for example), whereas discrete data will be exact (the score obtained by a single dart, for example).

● Frequency distributions

Sometimes the actual collection of data isn't very meaningful, and we wish to put the data into 'categories'. Take for example a list of student marks as percentages:

32, 78, 37, 65, 90, 87, 12, 41, 0, 91, 17, 65, 41, 45, 69, 54, 82, 65, 60, 51, 21, 37, 28, 53, 42, 48, 9, 71

Just looking at this doesn't really tell us much. It would be more useful to categorise the students into degree classifications. The fairly standard classifications in universities are the following:

- First class: 70 or more
- Upper second class: between 60 and 69
- Lower second class: between 50 and 59
- Third class: between 40 and 49
- Fail: less than 40

If you work through the list of students and count how many students fall into each class, you should get the following:

- First class: 6
- Upper second class: 5
- Lower second class: 3
- Third class: 5
- Fail: 9

Make sure you do this yourself. In lecture notes and books there are often many examples. If you don't get the answer you expected, then either you made a mistake (so do it again), or you didn't understand (so ask), or the lecturer/author made a mistake (this is possible, we all make mistakes): ask, and if they did, they will correct it – and they would like to know!

This is an example of a *frequency distribution*; instead of allocating the precise mark to each student, you are placing them in an appropriate category to get a more concise view of the results. Remember that seeing every single piece of data is often difficult to comprehend (refer back to the election example earlier), so summaries are often needed.

Often, in such cases, once a value has been placed in a category, we consider its value to be the midpoint of the range when we come to analyse it: this is something we shall return to later.

The categories are easy to create when you have discrete variables like the student marks, but it gets trickier with continuous variables, because you can have values that lie right on the edge of an interval. For example, you might want to class the speed of cars on a road into the following categories:

- less than 30 mph
- between 30 mph and 40 mph
- between 40 mph and 50 mph
- above 50 mph

The problem comes when a car is clocked exactly at one of the boundaries. Where do we put a car that is measured at exactly 40 mph? Should it go in the second or the third category? One obvious solution is simply to make the second category between 30 and 39 mph and the third category between 40 and 49 mph, etc. But then suppose we are measuring our speeds to 1 decimal place. Where does a car clocked at 39.5 mph fit? It goes into no categories at all now!

The solution here is to be more specific in what you mean by the intervals. Instead of 'between 30 mph and 40 mph' you could specify 'in the range from 30 mph to less than 40 mph', which you might abbreviate as '30 to <40' using the 'less than' symbol $<$. Writing our categories like this, we can define them as

- <30
- 30 to <40
- 40 to <50
- $\geqslant 50$

(note that we need the $\geqslant$ symbol for 'greater than or equal to' in the last one, to make sure that 50 is included in the range) and now every observation will fit into exactly one category.

You could write this in a different way; it's not really important how it is written as long as it is clear, and you have ensured that every value fits into exactly one category.

Summary

Statistics is essentially the study of data. It is used in a huge variety of areas; virtually any subject will need some element of data analysis and study. Remember that there are various aspects to statistics: the actual data collection, the presentation of the data (descriptive statistics), and the conclusions that can be drawn (inferential statistics). We shall of course explore this in much more detail as we go through the book.

Get yourself involved with a subject by realising how often it appears and plays a part in what you are doing. Statistics are everywhere; watch the news for half an hour, or read a newspaper, and you'll see numerous statistical figures presented to you. Think about your studies, and how there are 'average marks' and so on – all of this is statistics. Understanding why a subject is relevant and all around you makes it more real and then much easier to learn and understand.

Exercises

Every chapter contains exercises at the end. In most of the chapters these will be mathematical exercises to complete, but for this introductory chapter some of them are more open-ended.

1 In a given day or week, write down every aspect of statistics you see. This might be on TV or radio news, or in your studies. Every aspect counts – every average, diagram, chart, etc., and every figure released (inflation has risen by 0.1%, etc.). You might well surprise yourself by how much you actually see statistics!

2 Create a frequency distribution of the following recorded speeds of cars on a busy road, using the categories below.

29.8, 32.5, 39.9, 40.4, 28.3, 26.5, 30.0, 32.5, 40.0, 58.2, 28.0, 32.1, 31.4, 36.3, 39.2, 28.7, 31.1, 33.8, 29.8, 34.0, 31.4, 55.2, 40.1, 21.2, 21.4, 29.3, 32.7, 35.4, 33.0, 27.5, 22.3, 30.1, 30.0, 32.5

Categories:

    20 mph to <30 mph
    30 mph to <40 mph
    40 mph to <50 mph
    50 mph or more

Protesters are saying that far too many cars are driving dangerously
– significantly above the speed limit of 30 mph. Do the figures
support this view? (Just make an informal comment, looking at the
results. No need to do any calculations or anything – we shall do this
later.)

3 In a cricket match, a bowler's average is worked out by dividing
the number of runs conceded by the number of wickets taken. The
lower the average, the better.

- In the first match, Player X concedes 60 runs and takes 2 wickets.
  In the second match, they concede 72 runs and take 2 wickets.
- In the first match, Player Y concedes 29 runs and takes 1 wicket.
  In the second match, they concede 175 runs and take 5 wickets.

Calculate who has the better (lower) average in each of the two
matches.

If you combine the two matches together (so divide the total number
of runs by the total number of wickets over the two matches), which
player has the better (lower) average?

Do you notice anything unusual about your answers?

*(Aside: This is an example of something called Simpson's paradox; feel
free to research this.)*

Presentation of data

Using various types of graph and chart to illustrate data visually

In this chapter we are going to investigate some basic elements of data presentation. We shall look at ways in which collections of data can be presented in an appealing way to give us a visual indication or 'summary' of what the overall data is telling us.

Key topics

- Analysing data
- Plotting graphs
- Drawing charts

Key terms
line graph scatter graph correlation bar chart histogram
pie chart

● Line graphs

When you have a series of measurements taken over time, they are often represented as a *line graph*. This involves plotting all the points on a graph, and then connecting them together from one point to the next. You often see this sort of graph in the news – for example, showing the interest rate over the past few months or years. Here are the UK inflation rates for an 18-month period in 2009/10 (you can obtain these figures, along with numerous other data, from the Office for National Statistics).

Jan 09	Feb 09	Mar 09	Apr 09	May 09	Jun 09	Jul 09	Aug 09	Sep 09
3.0	3.2	2.9	2.3	2.2	1.8	1.8	1.6	1.1

Oct 09	Nov 09	Dec 09	Jan 10	Feb 10	Mar 10	Apr 10	May 10	Jun 10
1.5	1.9	2.9	3.5	3.0	3.4	3.7	3.4	3.2

To plot these on a graph, you need to create two *axes*, one going horizontally across (which will represent our months), and one going vertically up (which will represent the inflation rate). Note that you should label the axes clearly as to what corresponds to what. For each month, go up to the appropriate inflation rate and mark the point with a cross, or a solid point, or some other clear symbol. For example, to plot the first value, go to the label for Jan 09, and then go up until you reach 3.0, marking it in some way.

Once you have filled in all the values, connect them together with a series of straight lines, as in Figure 2.1. Your graph may well look different, because you will probably choose different scales, titles, labels and so on, but this will give you an indication of what it should look like.

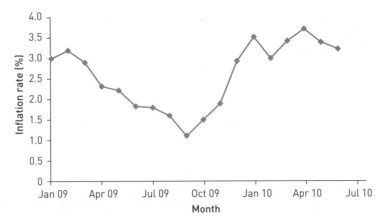

Figure 2.1 **Example of a line graph**

It's more obvious from this graph what the overall trend was, rather than just looking at a long list of numbers. You get the impression that the rate fell rapidly and then recovered, which is not as easy to tell from a simple, raw list of data and numbers.

For any set of data that is taken over a period of time at regular intervals, a line graph is usually a good way to show the overall trend.

● Scattergraphs

A scattergraph is what you get when you simply plot every point on the graph without making any attempt to connect them together. You can use it for data when both the x (horizontal) and y (vertical) axes of the graph can take a range of values (not necessarily in order of time – just a random collection of data). If the data is 'scattered' widely, then there appears to be no *correlation* (relationship) between the two things. If they form something close to a line, then there is a good correlation. For example, the following data might have been obtained by recording people's height (in metres) and their weight (in kg).

Height (m)	1.75	2.01	1.64	1.51	1.72	1.81	1.65	1.92	1.83	1.87
Weight (kg)	80	109	72	60	81	90	73	101	93	99

On a scattergraph this could be plotted as shown in Figure 2.2. Note the scale: why start at 1.4, not 0? There appears to be a strong correlation between height and weight: they seem to fall pretty well in a line. This is what you might expect: you would expect taller people to weigh more.

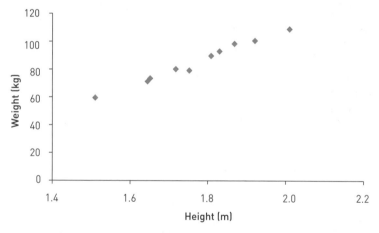

Figure 2.2 **Example of a scattergraph showing a positive correlation**

If the data lies in an 'upward line' like this (so that as one thing gets bigger, so does the other), you would say that there is a *positive correlation* between the two things. If the data slopes the other way (so that as one thing gets bigger, the other gets smaller) you would say that there is a *negative correlation*.

The words *strong* and *weak* are used for correlations. Where the two things are, clearly, very closely linked, you would say it is a *strong correlation*, whereas if there is some connection, but it is small, you would say it is a *weak correlation*.

Things aren't always correlated. Here is data analysing the same people with the number of shirts they own:

Height (m)	1.75	2.01	1.64	1.51	1.72	1.81	1.65	1.92	1.83	1.87
No. of shirts	5	3	14	8	2	10	9	20	1	7

Again draw a graph (Figure 2.3). This time there doesn't appear to be a connection: the points are scattered everywhere, and they don't lie in anything like a line. Again, this is probably what you would expect: why should a person's height relate to how many shirts they have? Hence there is no correlation here.

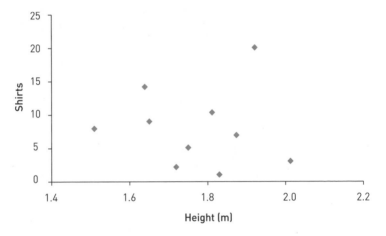

Figure 2.3 **Example of a scattergraph showing no correlation**

Also note that there would be no point in drawing a line between the points here in these sorts of graph. We use a line graph when we have a series of measurements taken over time, but for a general set of measurements with no time flow to them, we just use a scattergraph.

● Bar charts

When the data you get can be classified into a small number of 'categories', you can often usefully express it as a bar chart.

For example, suppose a survey was done of people's eye colour, and the following results were obtained:

Blue: 32 people
Brown: 78 people
Green: 6 people

To plot this as a graph, label the horizontal axis with 'blue', 'brown' and 'green', and then draw 'bars' for each one to represent the appropriate number. You should get a graph that looks something like Figure 2.4.

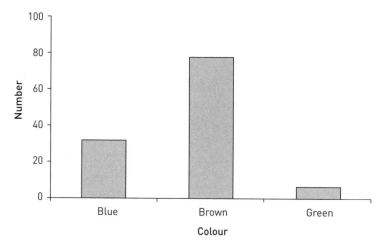

Figure 2.4 **Example of a bar chart**

You can do this in other ways as well. For example, you might put the colours on the vertical axis and draw horizontal bars, or you might colour-code the bars, or you might draw them without gaps between the bars. It doesn't really matter; choose the method of presentation that suits what you want to present the best.

The word *histogram* is often used instead of 'bar chart'. Although it does have a precise mathematical meaning, most people use it interchangeably for 'bar chart', and you can assume that a histogram is simply a bar chart as we have drawn here.

A pie chart is a way of representing proportions. For example, suppose 50 people were surveyed and asked their views on a product. The results were as follows:

Very good: 23
Good: 11
Average: 8
Poor: 6
Very poor: 2

To draw these figures on a pie chart, we are going to divide a circle into segments corresponding to the options: the larger the segment, the more people chose the option that it represents.

A circle is made up of 360 degrees (written as 360°). You should know how to use a protractor to measure an angle; if you don't know, then ask someone to show you how to use it.

> ✔ Do get hold of a basic maths set, with a ruler, protractor, set square, etc. You can buy them cheaply from standard stationery/discount stores. They can come in useful, even at times you don't expect; there are lots of times you just want a ruler to draw a straight line, for example.

The proportion of people who opted for 'very good' is $\frac{23}{50}$. A circle has 360° in it. The technique to follow is to convert this to the corresponding proportion of 360: this is the number of degrees that the 'very good' segment should occupy. Do this:

- Work out the proportion as a decimal on your calculator. In this case we get 23/50 = 0.46.
- Multiply this number by 360 on your calculator; you get 165.6.
- So the 'very good' segment should be 165.6°.

Similarly, you will find that the 'good' segment should be 79.2°, the 'average' segment should be 57.6°, the 'poor' segment should be 43.2°, and the 'very poor' segment should be 14.4°.

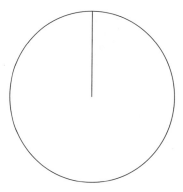

Figure 2.5 **Example of a pie chart showing the vertical radius only**

Draw a circle, and draw a vertical radius (a line from the centre to the edge) (Figure 2.5). Now, **carefully**, with a protractor draw the 'very good' segment, so it should form an angle of 165.6°. You get something like Figure 2.6.

very good

Figure 2.6 **Example of a pie chart showing the 'very good' segment only**

Continue like this for each segment. Label each segment clearly. When you are presenting a pie chart as part of a formal document, we recommend you use a different colour for each segment.

Sometimes in pie charts we would also include the numbers or, more commonly, the percentage of people making each choice, as in Figure 2.7 (you should check for yourself that the percentages are correct).

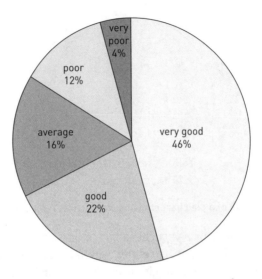

Figure 2.7 **Example of a pie chart showing all segments and percentages**

 ## Summary

It is vital to present data in an easy-to-understand way. When people see the results of a by-election, for example, most of them don't particularly want to see the actual number of votes cast for each candidate; they want them expressed in a visual format that makes the result jump out at them. Similarly, not everyone can see immediately what a series of inflation rate figures mean, but a picture shows the trend.

People don't want to understand long lists of figures; they want a simple picture summarising them. In a sense, this is what statistics is all about.

When presenting data, use the most appropriate form, but also try to catch the attention of the audience who are looking at it – whether this be in a report, or in a presentation, or in whatever form you are presenting your results. Good use of colours can help attract people, and by drawing attention to the main features you want understood, you can use the statistics to highlight clearly what you want to say. Of course, keep things simple, and don't confuse people, but good choices of colours and presentation methods can very effectively sell your message.

Exercises

1 (a) Present the following data showing the number of eggs laid on a chicken farm each day as a line graph:

Day	1	2	3	4	5	6	7	8	9	10
Eggs	46	52	53	60	61	24	25	30	32	37

(b) Present the following data showing the interest rate of a credit card over 12 months as a line graph:

Month	Jan	Feb	Mar	Apr	May	Jun	Jul	Aug	Sep	Oct	Nov	Dec
Rate	10.1	10.5	11.8	13.7	12.9	12.9	13.1	12.0	11.5	10.9	11.4	11.9

2 The following data shows the average temperature in successive weeks, and the corresponding figures for sales of drinks and batteries in a store. Plot the data on two scattergraphs: (a) one for temperature and drinks, and (b) one for temperature and batteries. Does there appear to be a correlation in either case?

Temperature	20	25	32	21	28	23	18	14	12	8
Drinks	120	220	300	115	250	170	120	40	45	10
Batteries	90	60	50	120	130	100	50	30	60	90

3 (a) In a survey of an online shop, the percentages of people rating the quality of the shopping experience in various categories were:

Excellent: 26%
Good: 43%
Average: 12%
Poor: 13%
Very poor: 6%

Plot this information as a bar chart and as a pie chart.

(b) In an election for a class representative, the 40 votes were cast as:

Alex: 3 votes
Bina: 19 votes
Chloe: 12 votes
Darren: 6 votes.

Plot this information as a bar graph and as a pie chart.

3 | Averages

Different methods of working out the 'average' or 'typical' value in a set of data

Given a set of data, it is nice to have some figure to give us the 'typical' value of all the values in the data set. There are various ways in which this can be defined – some more useful than others in different circumstances. In this chapter we shall look at some of the more common measures used to give us an 'average' value.

Key topics
- Mean, median and mode
- Geometric and harmonic mean

Key terms
average measure of central tendency mean median
mode geometric mean harmonic mean

An *average* (which goes by other names, such as a *measure of central tendency*, or a *measure of location*, and others) is essentially a way to work out what we consider to be the 'typical' value of a set of data. There are many ways in which this can be defined; we shall look at some of the most common.

Mean

When most people refer to an average, they are referring to what is known as the *mean*. This is easily calculated:

- Add up all the values in the data set.
- Divide by the number of values in the data set.

For example, suppose the marks in a test for a small class of five students were 2, 3, 6, 7 and 7. There are five values in this data set (five students), so you need to work out $\frac{2 + 3 + 6 + 7 + 7}{5} = \frac{25}{5} = 5$. So the mean is 5. Note that the mean might not be an actual value that any student obtained (as in this case – no-one actually obtained 5), but it gives an indication of the 'average' performance of the students: 'The average mark was 5.'

Note that often the mean will come out to be a decimal number, and so you may need to give the answer to an appropriate number of decimal places (or leave it as a fraction).

● Median

The mean is the most common measure of an 'average', but it is not always the most appropriate. Consider the following data, which shows the salaries of nine employees in a company (in thousands of pounds):

10   15   12   20   16   24   18   1000   19

From this data, it seems that most workers are on fairly standard salaries in the £10,000–£25,000 bracket, apart from one (presumably the boss) on £1,000,000. If you work out the mean of these salaries, you get

$$\frac{10 + 15 + 12 + 20 + 16 + 24 + 18 + 1000 + 19}{9} = \frac{1134}{9} = 126$$

and so the mean is 126: the 'average' salary is £126,000. Is this representative of the data? If this company advertised a vacancy and stated that the average salary was £126,000, wouldn't you feel somewhat disappointed when you joined the ranks on the sort of standard salary most people are earning?

The problem with the mean is that the one really huge salary is massively affecting the average. This happens quite a lot with salaries, and also with things such as house prices (think about how there are many cheap houses, but some houses worth millions and millions which really push up the mean).

The *median* is a way to try and address this, by giving the 'middle' value. The basic idea is that half the values are below the median, and half the values are above it. To do this, first list the values in ascending order:

10 12 15 16 18 19 20 24 1000

Now pick the value that lies in the middle, which is 18. So the median salary is £18,000. Do you think this is a more realistic measure of the 'average'?

> ✔ You can see from an example like this that statistics can be misleading. Next time you hear the word 'average' on the news, what sort of average do you think they are talking about? Is it ethical to use a misleading definition of 'average'? Think about these sort of things, to make topics relevant.

There is one issue with the median that we need to discuss. Suppose a new employee joins on a salary of £26,000, so the value is 26.

Then the values now (in order) are

10 12 15 16 18 19 20 24 26 1000

The problem is that there is no single 'middle' value here: both 18 and 19 are 'in the middle'. This is going to happen every time we have a data set with an even number of values. The convention in this case is to take the mean of the two middle values; we work out $\frac{18 + 19}{2} = 18.5$, and so the median value is 18.5 (corresponding to £18,500).

To summarise all of this, to work out the median:

- Put the values in ascending order.
- If there are an odd number of values, the median is simply the middle value.
- If there are an even number of values, take the mean of the two middle values (add them up and divide by 2) to give the median.

● Mode

A very simple, but not commonly used, method of deciding a 'typical' value is simply to count which value appears the most times in the data set. This is called the *mode*.

For example, in the set of data 2, 6, 5, 2, 3, 7, 2, 6, 5, the value 2 appears more than any other value – it appears three times, no other value appears more than twice – and so the mode is 2.

It is possible to have more than one mode. For example, in the data set 3, 4, 3, 5, 4, both 3 and 4 appear the most times (twice), and so there are two modes: 3 and 4. This is known as a *bimodal* data set ('bi' means 'two' – there are two modes).

In a case such as 1, 2, 7, 3, 5, 4, every value appears only once, so we would have to say that every value is the mode in this case.

Note that the mode is the only measure of 'average' that makes sense for data sets that do not consist of numbers. For example, the mode of the set of letters

a, b, e, b, c, d, b, e

is *b*, since it appears the most times.

This is a very basic concept of 'typical': it simply says that the typical value is the most common one, which is not usually very useful. For example, the mode of exam marks of eligible students will often be 0, as there will be a few students who do not take the exam, and some who have not revised at all, and do not get any marks: so 0 is the most commonly occurring mark, which is entirely unrepresentative of the performance of the students as a whole. Hence the mode is not widely used, unless you need to know the most common occurrence in a set of values, such as in an election – the winner is the mode of the votes.

These three – mean, median and mode – are the most common concepts of 'average' that you will see, but there are many others. Let us look at a couple of these.

● Geometric mean

The *mean* that we referred to earlier (the standard definition of 'average') is formally known as the *arithmetic mean*. The word *arithmetic* here refers to the fact that we add up the values. Instead of adding up the values, we could multiply them all together. This is what we do in a *geometric mean*. The word *geometric* refers to multiplying; if you know about arithmetic and geometric progressions, then you will understand the connection between the words 'arithmetic' and 'geometric', and between 'addition' and 'multiplication'.

When calculating the geometric mean, instead of dividing by n (where n is the number of values) we take the nth root of the product.

If you have forgotten what the nth root of a number is, I'll give a brief revision below, but if this is difficult for you, you should go back and revise it.

> The nth root of a number is another number, which gives the original number when you multiply n copies of it by themselves. This is clearer by examples:
>
> - The 2nd root (normally called the square root) of 9 is 3, since $3 \times 3 = 9$ (two 3s multiplied together make 9).
> - The 3rd root (normally called the cube root) of 8 is 2, since $2 \times 2 \times 2 = 8$ (three 2s multiplied together make 8).
> - The 5th root of 32 is 2, since $2 \times 2 \times 2 \times 2 \times 2 = 32$ (five 2s multiplied together make 32).
>
> The nth root of a number x is written $\sqrt[n]{x}$. So for example, $\sqrt[5]{32} = 2$, since 2 is the 5th root of 32, as in the example above. Note that, because square roots are so common, the convention is to omit the 2 in those roots and just write $\sqrt{x}$. So, for example, we would write $\sqrt{9} = 3$ rather than $\sqrt[2]{9} = 3$.

> ✔ If you have forgotten this maths, then you need to go back and revise it. Maths plays a huge role in statistics, and so if you are a bit unsure of it, now is the time to go back and refresh your memory; it will be time very well spent.

To work out the geometric mean of a data set:

- Multiply all the values together.
- Take the nth root of the answer, where n is the number of values.

You will need a calculator or computer to work out the nth root in all but the most simple of cases. Make sure you know how to work this out on your calculator or computer; if you don't know how, then find out: ask someone, read the manual, or search on the Internet.

Example

To work out the geometric mean of 2, 5 and 6, first multiply them together: $2 \times 5 \times 6 = 60$. There are three values, so we need to take the cube root (3rd root). Working out $\sqrt[3]{60}$ on a calculator or computer gives 3.915 (to 3 decimal places).

The geometric mean can be very useful. It is especially used in finance for things such as the average percentage return on investments over a number of years, where returns are calculated multiplicatively, and so the geometric mean provides a better measure of 'average' than the standard arithmetic mean.

● Harmonic mean

The harmonic mean is another way of measuring an 'average' that at first seems completely convoluted and pointless, but is actually useful. It was first developed and then used widely in ancient Greece, where it had applications to music: it can help with working out the appropriate length of strings on musical instruments, for example.

To work out the harmonic mean of a set of values, you need to do the following:

- Calculate the reciprocal of each value (the reciprocal of a number x is $1/x$, so for example the reciprocal of 4 is $\frac{1}{4}$).
- Add all these reciprocals together.
- Divide n (the number of values) by the sum of the reciprocals just worked out.

In essence, this is like doing a normal arithmetic mean, but using reciprocals everywhere. To illustrate by example:

Example

To work out the harmonic mean of the numbers 2, 5 and 6, first calculate their reciprocals and then add them together, giving $\frac{1}{2} + \frac{1}{5} + \frac{1}{6}$. Now divide n, which is 3 in this case (there are three values), by this sum. We need to work out $\dfrac{3}{\frac{1}{2} + \frac{1}{5} + \frac{1}{6}}$, which, if you work it out on a calculator or computer (be careful how you type it!) should give 3.462 to 3 decimal places. This is the harmonic mean.

The harmonic mean is surprisingly useful. Consider the following question:

A and B are two towns. I drive at 20 mph from A to B, and then at 30 mph back from B to A. What was the average speed during my overall drive?

Almost everyone would intuitively say the average must surely be 25 mph. But after some thought, you might start to think this might not be right. You are spending less time at the 30 mph speed (going faster, so less time to travel), so it's not such an easy calculation as you might think.

I won't prove it here (research it if you want to), but I can tell you that the average speed is given by the harmonic mean. There are two values (20 and 30), and so you need to work out $\dfrac{2}{\frac{1}{20} + \frac{1}{30}} = 24$ mph.

You might not intuitively believe this, and wonder why this could be the 'average'.

To illustrate, let's take an example: say A and B are 60 miles apart (chosen for convenience to make the answers easy, but it works for any distance). It takes you 3 hours to get there (driving at 20 mph) and 2 hours to get back (driving at 30 mph). So in total you have driven 120 miles in 5 hours, which is equivalent to 24 mph on average $\left(\dfrac{120}{5} = 24\right)$. So this is actually a sensible measure of 'average': your overall journey is equivalent to one in which you drove consistently at an 'average' 24 mph all the time.

● Other averages

There are many other types of average. You could do something as simple as just add up the smallest value and largest value and divide by 2 – sometimes referred to as the *midrange*. This is a very simplistic measure that ignores all but the two most extreme values. At the other end of the scale you could do very complicated calculations involving exponentials and logarithms: there are many possible definitions of 'average'!

In statistics, you will often see various definitions of concepts. Here we have seen different ways to define the concept of an 'average' value; later we shall see different tests, different distributions, and so on. Why do we need them? Quite simply, different situations have different needs, and all the definitions are useful in some situations. When you are learning statistics, bear in mind that the definitions are meant to be useful. People didn't just make them up for fun; they really are useful!

 Summary

The aim here is to give a measure of the 'typical' value of a set of data. But what we mean by 'typical' depends on our intention. Do we mean the most common value (the mode), the middle one (the median), the standard average (the mean), or perhaps something else useful in another scenario (the geometric mean in finance, or the harmonic mean in travel calculations, and so on)? The different definitions are all useful. Although in most cases the word 'average' is understood to refer to the arithmetic mean, this is not the only measure of a 'typical value', which I hope you have seen from this chapter.

 Exercises

Give any decimal answers to 3 decimal places.

1 Calculate the mean of the following sets of data:

Example: 3, 5, 2, 9, 1

Solution: Add up the values and then divide by 5 (the number of values in the data set) to get $\frac{3 + 5 + 2 + 9 + 1}{5} = \frac{20}{5} = 4$ and so the mean is 4.

(a) 2, 5, 10, 11
(b) 3, 4, 2, 1, 8, 5
(c) 1, 2, 3, 4, 5, 6, 7, 8, 9, 10
(d) 3, 3, 3, 3, 3

2 Calculate the median of the following sets of data:

Example: 9, 2, 1, 4, 5, 3, 8

Solution: First list the values in order, giving 1, 2, 3, 4, 5, 8, 9, and then take the middle element, which is 4: so the median is 4.

(a) 7, 2, 1, 9, 5
(b) 3, 6, 2, 8, 1000, 10
(c) 7, 2, 5, 3, 7, 9, 10
(d) 5, 5, 5, 5, 5

3 Calculate the mode of the following sets of data:

Example: 2, 5, 2, 9, 3, 6, 2, 5, 9

Solution: The most common value is 2 (it appears three times; nothing else appears more than twice) and so the mode is 2.

(a) 2, 7, 3, 3, 7, 5, 2, 8, 2, 1
(b) 4, 7, 1, 8, 1, 4, 9
(c) 3, 3, 3, 3, 3
(d) c, b, a, b, c, d, e, d, c

4 Calculate the geometric mean of the following sets of data:

Example: 2, 3, 4

Solution: Multiply the values together to get $2 \times 3 \times 4 = 24$, and then take the third root $\sqrt[3]{24}$, which gives (using a calculator) 2.884 to 3 decimal places.

(a) 2, 8
(b) 2, 5, 6
(c) 1, 2, 3, 4
(d) 4, 4, 4, 4, 4

5 Calculate the harmonic mean of the following sets of data:

Example: 2, 3, 4

Solution: You need to work out $\dfrac{3}{\frac{1}{2} + \frac{1}{3} + \frac{1}{4}}$ (there are 3 values), which gives (using a calculator) 2.769 to 3 decimal places.

(a) 1, 2
(b) 10, 20
(c) 1, 2, 3
(d) 6, 6, 6, 6

Cumulative frequencies and percentiles

Understanding cumulative frequencies and the use of percentiles to give measures to 'positions' in the data

In the last chapter we discussed the median, which was the 'middle value' in the sense that, essentially, half of the values are above the median, and half of the values are below it. Here we shall generalise this idea. First we need the idea of cumulative frequencies.

Key topics
- Cumulative frequencies
- Percentiles
- Quartiles

Key terms
cumulative frequency cumulative frequency graph percentile
upper and lower quartiles

● Cumulative frequencies

Consider the following frequency distribution, which shows the times of the goals scored in a weekend of football matches:

Time period	Number of goals
>0 to ≤10 minutes	5
>10 to ≤20 minutes	13
>20 to ≤30 minutes	4
>30 to ≤40 minutes	11
>40 to ≤50 minutes	19
>50 to ≤60 minutes	3
>60 to ≤70 minutes	11
>70 to ≤80 minutes	14
>80 to ≤90 minutes	24

Note that we are using the shorthand notation, similar to that discussed in Chapter 1. For example, the range '>40 to ≤50 minutes' includes every goal scored between 40 and 50 minutes, including 50 minutes exactly, but not including 40 minutes exactly; this goes in the previous range.

Note that there were 104 goals scored in total (add up the values in all of the categories).

The idea of a *cumulative frequency* is to keep track of how many values lie either in a particular range, or in any of the ranges smaller than it.

So, for example, the cumulative frequency of the second range is 18 (the 13 goals scored in that range, together with the 5 in the previous range). The cumulative frequency of the third range is 22 (the 4 goals scored in that range, together with the 18 you just worked out from the two previous ranges), and so on. This can be recorded by simply adding an extra column to the table. The easiest thing to do is just add the number of goals in each range to the cumulative frequency worked out 'so far'. You should get the following table:

Time period	Number of goals	Cumulative frequency
>0 to ≤10 minutes	5	5
>10 to ≤20 minutes	13	18
>20 to ≤30 minutes	4	22
>30 to ≤40 minutes	11	33
>40 to ≤50 minutes	19	52
>50 to ≤60 minutes	3	55
>60 to ≤70 minutes	11	66
>70 to ≤80 minutes	14	80
>80 to ≤90 minutes	24	104

Note that the final cumulative frequency is 104, which is what you would expect (there were 104 goals in total).

The cumulative frequency allows us to answer questions such as 'How many goals were scored in the first hour of matches?' This is all the goals scored in the first six ranges (up to 60 minutes), which is exactly the cumulative frequency - so we can simply read off that there were 55 goals scored in the first hour.

● Cumulative frequency graphs

Cumulative frequencies are often represented on a *cumulative frequency graph*, with the cumulative frequencies always plotted on the vertical axis. If you draw a graph of the cumulative frequencies, you should get something like the diagram in Figure 4.1.

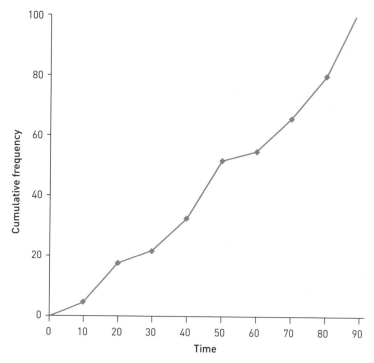

Figure 4.1 **Example of a cumulative frequency graph**

Note that we plot the points as a line graph with a succession of straight lines, and that we plot the points at the end of each range: so the first value is where the time is 10 (end of the range 0 to 10), and so on. This makes sense, as the lines connecting the point from one range to the next then cover the entire range.

We can use the graph to help us with such questions as 'How many goals were scored in the first half?' (which is 45 minutes). We can't answer this question exactly from the data; we know there were 33 goals in 40 minutes, and then 52 goals after 50 minutes, so the number at 45 must be somewhere between 33 and 52. This is all we know, but we can use the graph to estimate how many there 'probably' were at 45 minutes. Simply draw a vertical line from 45 minutes, and read off the corresponding value as best you can from the graph. If you do this, as in Figure 4.2, you should read off a value of around 42 or 43 goals.

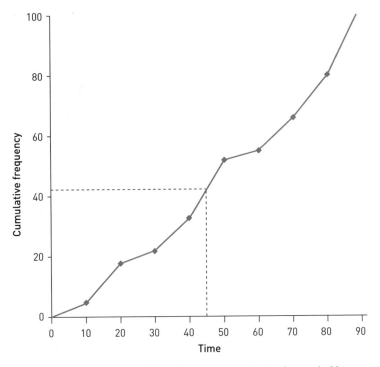

Figure 4.2 **Example of a cumulative frequency graph showing probable number of goals at 45 minutes**

This answer makes sense; 45 minutes is halfway through the range from 40 minutes to 50 minutes. There were 33 goals at the start of the range, and 52 at the end: halfway between these is 42.5, and so the 'likely' answer is around here, assuming the goals were evenly spread in the range. But remember, this is just a 'likely' answer. We have no way of knowing, from the data, the actual answer (all the goals might have been in the first few minutes of the range, before 45 minutes); all we can do is give the best estimate we can. This is a large part of statistical analysis!

● Percentiles

Recall that the median was defined as a 'middle' value: roughly speaking, half the values are below the median and the other half are above the median.

The concept of this can be generalised to *percentiles*. We define the nth percentile to be a number such that n% of the data values are below the percentile. So, for example, the 20th percentile is a value such that 20% of the data values are below it.

Unfortunately, defining formally how to calculate this without any confusion can be rather tricky, and there is no real standard definition. The sorts of problem that arise can be illustrated via the median. Intuitively, the median should be the 50th percentile (half of the values below the median), and it *almost* is. For example, in the set of data 2, 3, 7, 9, 9, 10 the median is 8 (the mean of the two middle values 7 and 9). This is fine; half of the numbers (3 of the 6 values) are less than the median 8. But consider a list with an odd number of values, such as 3, 5, 6, 8, 9. Then the median is 6 (the middle value), and it's not quite half of the values less than the median (only 2 out of 5 are less than it, and 2 out of 5 isn't half).

However, this is really a very minor point, and we shall gloss over it. There are ways to alter the definition slightly to make it more complicated, and reduce these sorts of problem slightly, but for the sort of large data sets for which we need and use statistics, the possible 'error' is so small as to make no real difference to us, and the fact that we usually have frequency distributions and/or sampling means that the values we have are not exact, and so we are estimating anyway.

> Remember that most statistical problems you will encounter will not be a complete set of precise data from the entire population. Because of samples, frequency distributions, etc., statistics can rarely give a precise answer; the idea is that it gives a 'likely', 'probable' or 'expected' answer.

Hence from now on we shall consider the median to be the same as the 50th percentile, and not worry about the very slight technical issue here.

As well as referring to the median as the 50th percentile, two other common percentiles are the 25th percentile and the 75th percentile. These are referred to as the *lower quartile* and *upper quartile* respectively: the names reflect the fact that they identify whereabouts the lower and upper quarter of the data values lie.

You can estimate these either graphically, or via a mathematical calculation.

● Calculating percentiles graphically

You can estimate percentiles via the cumulative frequency graph. Let us take the same data set as above (the times of goals in football matches) and try to work out approximately the upper and lower quartiles, and the median. Remember that there were 104 goals in total.

- The lower quartile is the 25th percentile. If there are 104 goals in total, then the 25% point of this is 25% of $104 = \frac{25}{100} \times 104 = 26$ (remember that a percentage is just a fraction out of 100), and so the lower quartile is the 26th goal scored.

- The median is the 50th percentile. If there are 104 goals in total, then the 50% point of this is 50% of $104 = \frac{50}{100} \times 104 = 52$, and so the median is the 52nd goal scored.

- The upper quartile is the 75th percentile. If there are 104 goals in total, then the 75% point of this is 75% of $104 = \frac{75}{100} \times 104 = 78$, and so the upper quartile is the 78th goal scored.

Now look at the cumulative frequency graph (Figure 4.3) and identify the times corresponding to the 26th, 52nd and 78th goals on the cumulative frequency axis.

Reading off from the graph, the lower quartile appears to be around 34 minutes, the median seems to be around 50 minutes, and the upper quartile seems to be around 79 minutes.

This method is fine, but it does mean you have to go to the trouble of drawing the graph, then drawing accurate lines and reading off the values as best you can. It is always better (to help cut out errors) to do things mathematically, when possible; can we do this here?

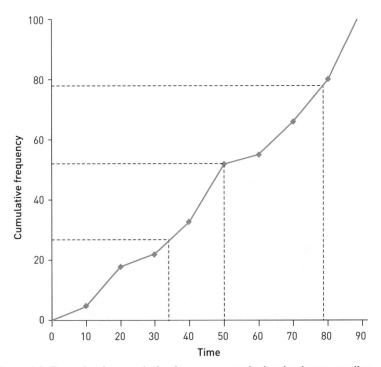

Figure 4.3 **Example of a cumulative frequency graph showing lower, median and upper quartile percentiles**

● Calculating percentiles via direct calculation

We can do this calculation directly without a graph. In fact it is exactly the same technique, but expressed mathematically.

Just as above, we can work out that we need to estimate the times of the 26th, 52nd and 78th goals.

Note that the 26th goal lies in the range from 30 to 40 minutes (there were 22 goals by 30 minutes, and 33 goals by 40 minutes, so the 26th goal must have been scored in this range).

The 26th goal is the fourth to be scored in this range (we have already had 22 goals, so the 26th is the fourth goal in this period (26 − 22 = 4), and there were 11 goals scored in total (total goals at the end is 33, and total goals at the start is 22, so 33 − 22 = 11 goals were scored in this range). So, assuming the goals were spread out equally over the range (which is a fair assumption to make in the lack of any other knowledge), you would expect the goal to have been scored 4/11 of the way through the range: that is, at position $30 + \frac{4}{11} \times (40 - 30)$. That is, starting from 30 minutes (start of the range), we want to move forward 4/11 of the range, which is of length $40 - 30$ (end of the range minus the start of the range) which is 10 minutes. This works out to be around 33.6 on a calculator, and so this is our lower quartile.

This can be visualised in the diagram in Figure 4.4, where the arrow is 4/11 of the way along the line from 30 to 40.

Figure 4.4 Example of linear interpolation

This is an example of something called *linear interpolation*. We shall see this again later; it's basically just estimating the position of something, given that we know it lies inside a range with two fixed endpoints.

This technique can lead to a general formula:

> To calculate the position of the *n*th value in a range from *a* to *b*, with starting cumulative frequency *s* and final cumulative frequency *f*, the formula is:
>
> $$a + \frac{n - s}{f - s}(b - a)$$

Check this formula with our example above: we have $n = 26$, $a = 30$, $b = 40$, $s = 22$ and $f = 33$, so we work out

$$30 + \frac{26 - 22}{33 - 22} \times (40 - 30) = 30 + \frac{4}{11} \times 10$$

just as above.

We can now work out the other two values in the same way, using this formula:

- The median (50th percentile) is when the 52nd goal was scored. This is in the 40 to 50 minute range, which started with 33 and ended with 52. So we have $n = 52$, $a = 40$, $b = 50$, $s = 33$ and $f = 52$. Putting these values into the formula, we get $40 + \frac{52 - 33}{52 - 33}$ $\times (50 - 40) = 40 + \frac{19}{19} \times 10 = 40 + 1 \times 10 = 40 + 10 = 50$. So the median is 50 minutes.

- The upper quartile (75th percentile) is when the 78th goal was scored. This is in the 70 to 80 minute range, which started with 66 and ended with 80. So we have $n = 78$, $a = 70$, $b = 80$, $s = 66$ and $f = 80$. Putting these values into the formula, we get
$$70 + \frac{78 - 66}{80 - 66} \times (80 - 70) = 70 + \frac{12}{14} \times 10 = 78.6 \text{ approximately}$$
(using a calculator). So the upper quartile is around 78.6 minutes.

Note that these answers match the answers we obtained graphically.

Note that the calculation for the median is to be expected at 50 minutes, since the cumulative frequency at the end of the 50 minutes is exactly the 52nd (middle goal). Also note that the answers all seem sensible (and match the graph you drew). If you can, check when you use a formula that the answers you get are realistic; if you get something that 'looks right' it gives you confidence in the formula, and that you have used it correctly.

Summary

Cumulative frequencies and percentiles are important tools. We shall see in the next chapter that quartiles are used to measure the spread of data. The concept of a percentile is quite important; it is useful to be able to identify a point before which, say 90% of the data lies. Again, we'll see this see this sort of idea later, when we deal with confidence limits. Something like a cumulative frequency doesn't seem that important when you first do it, but it leads into many other areas: this is true for most of what you do!

Exercises

For each of the sets of data below:

(a) Create a cumulative frequency table.

(b) Draw a cumulative frequency graph.

(c) Calculate, using the percentile approach as discussed in the chapter, the lower quartile, median, and upper quartile.

1 The speed of cars along a busy street:

Speed	Number of cars
>0 to ≤10 mph	2
>10 to ≤20 mph	3
>20 to ≤30 mph	5
>30 to ≤40 mph	28
>40 to ≤50 mph	56
>50 to ≤60 mph	16
>60 to ≤70 mph	9
>70 to ≤80 mph	1

2 The score obtained by a darts player with one dart:

Score	Number of scores
>0 to ≤10	8
>10 to ≤20	38
>20 to ≤30	17
>30 to ≤40	5
>40 to ≤50	12
>50 to ≤60	20

3 The marks obtained in a class test:

Mark	Number of students
>0 to ≤10	4
>10 to ≤20	12
>20 to ≤30	14
>30 to ≤40	8
>40 to ≤50	52
>50 to ≤60	32
>60 to ≤70	28
>70 to ≤80	40
>80 to ≤90	8
>90 to ≤100	2

Different methods of working out the 'spread' of a set of data

Averages are all about giving a 'typical' value to a set of data. But as well as knowing the 'typical' value, it is also important to know how spread out the data is: are all the values close together, or are they really spread out?

Key topics
● Range
● Variance
● Standard deviation

Key terms
range interquartile range variance population standard deviation
sample standard deviation

The *spread* (also referred to as a *measure of dispersion*) of a data set is intended to give us some idea as to how 'spread out' the data is. For example, in the data set of marks 48, 48, 50, 52, 52 the average is 50, and all the marks are very close to this average. However, with the set of marks 0, 10, 60, 80, 100 the average is also 50 (check this), but the marks are much more spread out, and far away from the average. In this chapter we shall look at some ways to measure the spread of data.

● A note on mathematical notation

In this chapter I have deliberately avoided using complex mathematical formulae. Later in the book I shall discuss sigma notation and present formulae for the sort of things we shall discuss here, but that is for later. For now, I want you to focus on understanding how to perform the calculations, and the basic ideas as to how they work.

✔ It is far more important to understand how to do something than it is to learn a complex formula. If you understand how to do something, then learning the formula is easy enough, since you know what to do. If you just learn a formula but don't know what you are doing, then it is essentially useless. Focus your studies on learning *how* to do things in practice, not on 'rote learning' of formulae.

● Range

The simplest measure of spread is simply to take the difference between the smallest and largest values: this is known as the *range*. So with the data set 48, 48, 50, 52, 52 the range is $52 - 48 = 4$, and with the data set 0, 10, 60, 80, 100 the range is $100 - 0 = 100$. The smaller the range, the less the data is spread.

The problem with this is that only the two extreme values play any part in the measure of the spread. The spread of the data set 10, 10, 10, 10, 10, 100 is obviously different from the spread of the data set 10, 30, 50, 70, 90, 100, but they both have the same range (90). Also, as the range uses only the extreme values, one unusual value can make a very misleading measure of the spread. With the data set 2, 5, 3, 4, 5, 3, 2, 4, 10000, 5, 3 almost all the values are very close together, apart from one extreme value of 10,000. The range would simply record the spread as $10,000 - 2 = 9998$, and this one extreme value means that the range has totally misrepresented the spread of the data.

In such cases the extreme value might be an error (if this was a set of scientific experimental data, say) or just a particular exception (the multi-millionaire head of a company, say). One way around this is to use quartiles, as discussed in Chapter 4. The *interquartile range* is simply the difference between the upper and lower quartiles, and so essentially is the range of the middle 50% of the data. But although this does deal with one extreme value, it still does not really capture the actual spread, because it does not take into account all the values, in exactly the same way that the range fails to do.

● Variance

The *variance* is a measure of spread that, unlike the range, takes into account all the values in the data set. The essential idea is to measure how far each data value is away from the average, and then take the average of these distances. This means that every element of the data set plays a role. However, there are a couple of hurdles to overcome before we can provide the definition.

First, we don't want any negative differences. If the average was 2 and we had one value of 1 and another of 3, we wouldn't want to consider the differences as -1 and $+1$, or they would cancel out to give 0, and we wouldn't be getting a meaningful measure of spread. Also, we'd like values 'far away' to be given more weight; a value close to the mean shouldn't contribute much to the spread, but a value far away should count much more.

We can address both of these concerns by taking the *square* of the differences (i.e. multiply them by themselves). You should know that square numbers are always positive, so we don't hit any problems with negative numbers. Also, if you take the square of the numbers, then the further away they are, the more they contribute. Something at a distance of 1 from the mean contributes $1^2 = 1$ to the variance. Something at distance 2 contributes $2^2 = 4$, something at distance 3 contributes $3^2 = 9$, and so on: the further away the values are from the mean, the more they contribute.

Hence we are now in a position to make a formal definition of variance. After the definition, we shall illustrate with the two examples mentioned earlier.

To calculate the variance of a data set:

- Calculate the mean.
- Calculate the difference between each value and the mean.
- Square each of these values.
- Take the mean of these 'differences squared'.

Example 1:

The data set 48, 48, 50, 52, 52. Follow the steps precisely:

- Calculate the mean of these values. The mean is $\dfrac{48 + 48 + 50 + 52 + 52}{5} = 50$

- Now calculate the difference between each value and the mean (50). In the following I have included the minus signs and simply subtracted the mean from the values. You don't have to; you can just record the positive values if you like (the minus signs will go away when we square the numbers).

Value	48	48	50	52	52
Difference from mean	−2	−2	0	2	2

- Now square each of these values. You get the following table:

Value	48	48	50	52	52
Difference from mean	−2	−2	0	2	2
Difference squared	4	4	0	4	4

- Now take the mean of these 'differences squared'. That means add them up and divide by the number of values (5). You get $\dfrac{4 + 4 + 0 + 4 + 4}{5} = \dfrac{16}{5} = 3.2$.

Hence the variance of this data is 3.2, which is quite small. Now try this with the other data set we considered, which is much more spread out.

Example 2:

The data set 0, 10, 60, 80, 100. Follow the steps precisely:

- Calculate the mean of these values. The mean is $\dfrac{0 + 10 + 60 + 80 + 100}{5} = 50$.

- Now calculate the difference between each value and the mean (50). Again, in the following I have included the minus signs and simply subtracted the mean from the values. Remember that you don't have to; you can just record the positive values if you like (the minus signs will go away when we square the numbers).

Value	0	10	60	80	100
Difference from mean	−50	−40	10	30	50

- Now square each of these values. You get the following table:

Value	0	10	60	80	100
Difference from mean	−50	−40	10	30	50
Difference squared	2500	1600	100	900	2500

- Now take the mean of these 'differences squared'. That means add them up and divide by the number of values (5). You get $\dfrac{2500 + 1600 + 100 + 900 + 2500}{5} = \dfrac{7600}{5} = 1520.$

So the variance of the first set of data is 3.2 (quite small), and the variance of the second set of data is 1520 (pretty big). So the second set of data appears to be much more spread out that the first set of data; the larger the variance, the more 'spread out' the data is.

Although this takes longer to do than the simplistic idea of a range, it does take all the values into account, and is a much more accurate measure.

● Standard deviation

When calculating the variance, we used the square of the values, and we discussed the benefits of doing so. However, you might think that if we are squaring everything, then really we ought to finish by taking the square root of our answer to 'get it back to normal', since the square root is the opposite of squaring. The *standard deviation* is defined in this way, and is nothing more than the square root of the variance.

The *standard deviation* of a set of data is the square root of the variance.

So, for the two examples we discussed above:

- We calculated the variance of the data set 48, 48, 50, 52, 52 to be 3.2, and so the standard variation is $\sqrt{3.2} = 1.789$ (to 3 decimal places).
- We calculated the variance of the data set 0, 10, 60, 80, 100 to be 1520, and so the standard variation is $\sqrt{1520} = 38.987$ (to 3 decimal places).

And that's all there is to standard deviation – well, almost...

● Population and sample variance and standard deviation

The calculations given above are used when we know the entire data set. Because of this, they are often referred to as the *population variance* and *population standard deviation*; the word 'population' refers to the fact that we know every data element ('everything in our population').

But, as discussed previously, we often 'sample' from a set as in sampling a representative selection of the population for an opinion poll, for example. In this case we have only a sample, rather than the entire population, to work with.

Calculating the variance and standard deviation of a sample is almost identical to working with the entire population, with one small difference. We work out the differences squared in exactly the same way. The only change comes at the next step. Instead of working out the mean (adding up the square values, and then dividing by n, where n is the number of values), we divide instead by $n - 1$. This is the only change; we still work out the sample standard deviation as the square root of the number we obtain.

To illustrate this, consider again the data set 48, 48, 50, 52, 52, but this time imagine that it was only a sample of values from a large data set. The differences squared are worked out just as before:

Value	48	48	50	52	52
Difference from mean	−2	−2	0	2	2
Difference squared	4	4	0	4	4

Now add these 'differences squared' and divide, not by the number of values (5), but by 1 less than this, so divide by 4. You get $\dfrac{4 + 4 + 0 + 4 + 4}{4} = \dfrac{16}{4} = 4$ as the sample variance, and so the sample standard deviation is $\sqrt{4} = 2$.

The calculation is almost identical, apart from the use of $n - 1$. Why do we do this? It is beyond the scope of this book to explain why. It has to do with the fact that using $n - 1$ can be shown mathematically to be the 'optimum' (best) number to divide by, to take into account the fact that we are dealing only with a sample, and not with the full population; essentially, subtracting the 1 accounts for the uncertainty in the sample, and for the fact that it is not fully representative. This takes a long, complex, mathematical proof, which I shall not confuse you with here, although you are welcome to research it if you wish.

Occasionally you have to learn something for which the explanation is really too complex for the level you are currently at. In cases like this, at least learn some sort of reason for it, even if the full reasoning is beyond what you need to know. In this case the $n - 1$ is used because we have only a sample and not all of the data, and $n - 1$ works better in this case; the -1 accounts for the uncertainty in the sample. It will help you remember things if you have some sort of idea in your head, and you can also research for yourself to find out more if you are interested.

 Summary

As we have seen, just giving an 'average' is not enough to represent a set of data; two wildly different sets of data can have the same average. We also need to make a measure of 'spread'. How spread out are the values? Are they all close to the average, or are they really spread out, with many far away? Calculating the standard deviation is a very common method for expressing the spread of data. You will see it often, and you need to know how to calculate it. Just make sure you know whether you are dealing with a population (the entire data set) or a sample (just a sample set of data from a large set).

 Exercises

1 Calculate the range of the following sets of data.

Example: 4, 7, 2, 5, 10, 6

Solution: The range is the difference between the largest and smallest values, so the range is $10 - 2 = 8$

(a) 1, 2, 3, 4, 5

(b) 10, 3, 7, 20, 9, 1, 18

(c) 2, 3, 2, 3, 2, 100, 2, 3

(d) 5, 5, 5, 5, 5

2 Calculate the population standard deviation of the following sets of data.

Example: 4, 6, 10, 12

Solution: First calculate the mean. The mean is $\dfrac{4 + 6 + 10 + 12}{4} = 8$. Now create the table of differences squared:

Value	4	6	10	12
Difference from mean	−4	−2	2	4
Difference squared	16	4	4	16

Now add up the 'differences squared' and divide by the number of values (4) to get $\dfrac{16 + 4 + 4 + 16}{4} = 10$: this is the population variance. To get the population standard deviation, you need to take the square root of this, so the population standard deviation is $\sqrt{10} = 3.162$ (to 3 decimal places).

(a) 3, 4, 9, 12   (b) 40, 40, 50, 60, 60, 80

(c) 3, 5, 10, 91, 93, 95   (d) 5, 5, 5, 5, 5

3 Repeat Q2 but instead assume that this is only sample data, so instead calculate the sample standard deviation.

Example: 4, 6, 10, 12

Solution: We proceed almost exactly as in the previous question. First calculate the mean. The mean is $\dfrac{4 + 6 + 10 + 12}{4} = 8$. Now create the table of differences squared:

Value	4	6	10	12
Difference from mean	−4	−2	2	4
Difference squared	16	4	4	16

Now add up the 'differences squared' and divide by $n - 1$ (one less than the number of values – this is the only difference in the calculation to Q2). So we divide by $4 - 1 = 3$ to get $\dfrac{16 + 4 + 4 + 16}{3} = \dfrac{40}{3} = 13.333$ (to 3 decimal places): this is the sample variance. To get the sample standard deviation, you need to take the square root of this, so the sample standard deviation is $\sqrt{\dfrac{40}{3}} = 3.651$ (to 3 decimal places).

(a) – (d) as in Q2

4 The following marks (as percentages) were obtained in a class test by two groups of students.

Group A: 40, 50, 50, 60, 40, 60
Group B: 0, 70, 74, 70, 80, 0

(a) Which group had the higher mean average?

(b) In which group was the spread of marks smaller? (Use the population standard deviation.)

(c) If the pass mark is 40%, which group had the best pass rate (the highest percentage of students passing)?

(d) On the basis that 40% or above is a pass mark, and 70% or above is a distinction mark, which group do you consider was the 'best' of the two groups? On what basis do you draw your conclusion?

Working with frequency distributions

How to use statistical concepts with frequency distributions

In previous chapters we have discussed such things as presenting data, averages and spreads, where the data is specified as particular values. What happens, though, if the data is grouped into a frequency distribution? How do we calculate averages and spreads?

Key topics

● Presenting frequency distributions
● Averages and spreads with frequency distributions

Key terms

bar charts averages and spread with a frequency distribution
midpoint

In this brief chapter we shall look at how to use the techniques already discussed when our data is given as a frequency distribution – so, instead of actual data values, we are given the number of data elements that fall into a particular range.

For consistency, I am going to use the same frequency distribution as before to illustrate the ideas. This was goals scored in football matches:

Time period	Number of goals
>0 to ≤10 minutes	5
>10 to ≤20 minutes	13
>20 to ≤30 minutes	4
>30 to ≤40 minutes	11
>40 to ≤50 minutes	19
>50 to ≤60 minutes	3
>60 to ≤70 minutes	11
>70 to ≤80 minutes	14
>80 to ≤90 minutes	24

● Presenting frequency distributions with bar charts

The obvious way to present this sort of data graphically is to use bar charts. These are an entirely natural choice, since each bar can represent a range.

If you draw a bar chart of the data, you get something like the chart given in Figure 6.1 (draw this for yourself, carefully). You could, of course, present this data in other ways, but it does seem quite natural to use a bar chart here.

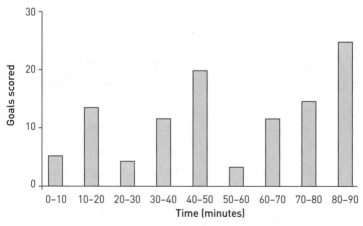

Figure 6.1 **Example of frequency distributions presented as a bar chart**

● Midpoints

Given a range, the *midpoint* is the point exactly halfway along the range. So, for example, in the range from 10 to 20 minutes, the midpoint is 15 minutes.

This is relevant when we consider our frequency distributions. For example, there are 19 goals in the 40–50 minute range. We have no way of knowing exactly when these goals were scored. They might all have been early in the range, or all have been later. The best we can realistically do is say that, on average, they will tend towards the middle of the range: some will be towards the start, some towards the end, and some in the middle, but overall we can expect them to average around the midpoint.

So, when we deal with calculations on frequency distributions, the logical thing to do is to assume every value has the midpoint. Of course, this is unlikely to actually be correct, but it's the best we can do. Of the 19 goals scored between 40 and 50 minutes, if we are going to guess a value for them, it makes more sense to assume they were all scored at 45 minutes than it does to assume they were all scored at 40 minutes, or 50 minutes, since we would expect 45 minutes to be the average overall.

We shall discuss this sort of idea in much more detail later when we consider distributions, but for now it makes sense to simply assume that all the values in a range take the value of the midpoint, so that our calculations should give us something reasonable. This allows us to give approximations for things such as the mean and standard deviation.

● Calculating the mean with frequency distributions

If we stick with the concept of assuming that all the values in a range take the midpoint, then we can calculate an estimate of the mean.

We just need to add the values and divide by the number of values as usual. To add the values, we just work through the ranges:

- There are 5 goals in the 0–10 minute range. Following the concept above, simply assume these all have value 5 (the midpoint of the range). So we have $5 \times 5 = 25$ contribution to the mean.

- There are 13 goals in the 10–20 minute range. Again following the concept above, simply assume these all have value 15 (the midpoint of the range). So we have $13 \times 15 = 195$ contribution to the mean.

- Similarly, there are 4 goals in the 20–30 minute range. Again following the concept above, simply assume these all have value 25 (the midpoint of the range). So we have $4 \times 25 = 100$ contribution to the mean.

- ... and so on through all the ranges.

By doing this, you are effectively adding together all the goal times (using our midpoint approximation for each range).

Therefore, to work out the overall mean, we have to add all of these together and divide by the number of goals (104):

$$\frac{5 \times 5 + 13 \times 15 + 4 \times 25 + 11 \times 35 + 19 \times 45 + 3 \times 55 + 11 \times 65 + 14 \times 75 + 24 \times 85}{104}$$

If you key this very carefully into a calculator, it should give the answer 53.17 (to 2 decimal places).

It is very easy to make a mistake when keying this into a calculator; be very careful when you do such calculations. It is always worth checking that your final answer is sensible; 53 minutes does seem about right for the mean. If you get a negative answer, or something more than 90 minutes, it must be wrong. Make sure you can use your calculator quickly and effectively, and that your answers are sensible. If you get an answer that looks 'about right' you can be confident that you've probably done the right thing.

Remember that this is only an approximation, because our goals are given only in ranges, not exactly in minutes, but it's as good an approximation as we can get with the data we have been given.

● Calculating the standard deviation with frequency distributions

You can calculate the standard deviation in exactly the same way: assume every value in the range takes the value of the midpoint, and do the usual calculation of standard deviation, as discussed in Chapter 5.

However, this is a really long calculation, and is much better left to a computer to do. If you try to do it yourself, you'll almost certainly make some mistakes, but go ahead and try if you feel brave. (The answer is 25.157 to 3 decimal places, assuming you work out the population standard deviation.)

Many statistical calculations are too tedious to bother doing by hand or on a calculator. This is where computer software such as Microsoft Excel can really help. This book is not intended as a resource to teach you a software package; it's intended to give you the basics of statistical theory. If you want to use a computer package, find a suitable book to deal with these advanced calculations in the software package of your choice!

Summary

All we have really done in this section is make a broad statement that if we don't know the exact value of a piece of data in a range, then we can just assume it lies in the middle. This is admittedly naïve, and certainly isn't perfect, but it does make sense to do it, and seems better than any other option.

Once we make this assumption, then we can use the ideas we already have, and everything follows smoothly. Statistics rarely gives perfect answers, but we can use it to give the best answer we can get.

Exercises

For each of the following probability distributions (these are the same ones as used in Chapter 4):

(a) Draw a bar chart of the data.
(b) Calculate an estimate for the mean.

1 The speed of cars along a busy street:

Speed	Number of cars
>0 to ≤10 mph	2
>10 to ≤20 mph	3
>20 to ≤30 mph	5
>30 to ≤40 mph	28
>40 to ≤50 mph	56
>50 to ≤60 mph	16
>60 to ≤70 mph	9
>70 to ≤80 mph	1

2 The score obtained by a darts player with one dart:

Score	Number of scores
>0 to ≤10	8
>10 to ≤20	38
>20 to ≤30	17
>30 to ≤40	5
>40 to ≤50	12
>50 to ≤60	20

3 The marks obtained in a class test:

Marks	Number of students
>0 to ≤10	4
>10 to ≤20	12
>20 to ≤30	14
>30 to ≤40	8
>40 to ≤50	52
>50 to ≤60	32
>60 to ≤70	28
>70 to ≤80	40
>80 to ≤90	8
>90 to ≤100	2

ESSENTIAL
MATHEMATICS

Factorials, permutations and combinations

Mathematical techniques for counting and to work out the number of ways of choosing elements from a set of data

Often we are faced in life with the problem of choosing a certain number of elements from a data set – for example, choosing the top three competitors in a talent competition. How many different possibilities are there for our choice? In this chapter we shall look at this idea, and introduce mathematical ways to count the number of ways in which we can make choices.

Key topics
- Factorials
- Permutations
- Combinations

Key terms
choosing factorial ordering permutation combination ^nP_r ^nC_r

Suppose you have five people in a competition. In how many ways can you order these people in a ranking from 1 to 5? In how many ways can you choose a winner and a runner-up, with everyone else left empty-handed? In how many ways can you pick three people to progress to a further round, regardless of order? These sorts of question can be answered using the mathematical concepts of *factorials*, *permutations* and *combinations*.

Factorials

Mathematically, the *factorial* of a positive integer is defined as the number you obtain by multiplying together all the positive integers up to and including the number in question. The factorial symbol is ! (an exclamation mark).

Formally, the definition of $n!$ for a positive integer n is

$$n! = n \times (n - 1) \times (n - 2) \times ... \times 3 \times 2 \times 1$$

So, for example, $4! = 4 \times 3 \times 2 \times 1 = 24$, and $5! = 5 \times 4 \times 3 \times 2 \times 1 = 120$.

You cannot sensibly define factorials for negative and non-integer numbers (why not?), although we make an exception for $0!$, which we define to be 1. This is done mainly for convenience (you will see a reason for it later). We also define $1!$ to be 1, which makes perfect sense.

You should have a factorial button on your calculator (probably labelled $n!$); find it and use it to check the answers for $5!$ and $4!$ given above.

Why is this relevant to the ideas of choosing that we discussed above?

Let us take the idea of five people in a competition; say they are called A, B, C, D and E. In how many ways can we order the five people (as first, second, third, fourth and fifth)? For example, we could order them (from first to fifth) as C, E, A, B, D, or D, B, A, E, C, and you can doubtless write down many more. How many possibilities are there? There are various ways to look at this, but the most natural for our purpose is the following way.

First we pick a winner. There are 5 possibilities for the first-place person (there are 5 people). For each of these possibilities there are 4 possibilities for the second place person (any one of the 4 left, after the first place is chosen). Then for each of these possibilities there are 3 possibilities for third place (any one of the 3 left, after the first and second places are chosen). Then for each of these there are 2 possibilities for the fourth place (any one of the 2 left, after the first, second and third places are chosen). Finally there is only 1 choice left for fifth place, since everyone else has been allocated, and so we have only the last remaining person to go fifth.

Looking at this mathematically, for each of the 5 possibilities for first place there are 4 possibilities for second place, and so we have 5×4 possibilities so far. Then for each of these there are 3 possibilities for third place, and so on. So we end up with a total of 5 × 4 × 3 × 2 × 1 possibilities.

But this is exactly what a factorial is: the number of possibilities is 5! = 5 × 4 × 3 × 2 × 1 = 120, and so there are 120 possible ways to order these 5 people. This is a much easier way to get the answer than trying to list every possibility on paper! This works in general:

> The number of possible ways to order a set of n elements is $n!$

A particular 'order' (such as C, E, A, B, D or D, B, A, E, C in this case) is called an *ordering*.

● Permutations

We discussed above why the number of possible orders of n elements is $n!$. This can be generalised. For example, how many ways are there to choose the winner and runner-up from our 5 people (the rest are not given a place)?

There are 5 possibilities for the winner, and then for each of these 4 possibilities for the runner-up, and so in total there are 5 × 4 = 20 possibilities. (If you were to write them all down, the 20 possibilities would be AB, AC, AD, AE, BA, BC, BD, BE, CA, CB, CD, CE, DA, DB, DC, DE, EA, EB, EC, ED.)

How many ways are there to choose first, second and third places, to award gold, silver and bronze medals? There are 5 possibilities for the gold medal. For each possibility, there are then 4 possibilities for the silver medal, and then for each of these, 3 possibilities for the bronze medal. So in total there are 5 × 4 × 3 = 60 different possibilities for the order of the three medals (I won't write these all down!).

Suppose instead there are 10 people. How many ways are there to choose first, second and third? The number of possibilities is 10 × 9 × 8 = 720. Similarly, how many ways are there to choose first, second, third, fourth and fifth? This would be 10 × 9 × 8 × 7 × 6 = 30,240.

Hopefully you are starting to see the pattern:

- The number of ways to order 2 elements from a set of 5 elements is 5×4.
- The number of ways to order 3 elements from a set of 5 elements is $5 \times 4 \times 3$.
- The number of ways to order 3 elements from a set of 10 elements is $10 \times 9 \times 8$.
- The number of ways to order 5 elements from a set of 10 elements is $10 \times 9 \times 8 \times 7 \times 6$.

In general, to order r elements from a set of n elements, you need to work out $n \times (n-1) \times \ldots \times (n - r + 1)$. Check for yourself that this formula is accurate for the four examples above. For example, in the last one, $n = 10$ (a set of 10 elements) and $r = 5$ (ordering 5 elements), so $n - r + 1 = 10 - 5 + 1 = 6$, and so we need to work out $10 \times 9 \times 8 \times 7 \times 6$.

A way of ordering the elements is called a *permutation*, and the total number of possibilities is called the *number of permutations*.

To summarise:

> The number of permutations of r elements from a set of n elements is $n \times (n - 1) \times \ldots \times (n - r + 1)$.

This is normally given in a slightly different form, purely in terms of factorials. It probably isn't immediately obvious to you, but $n \times (n - 1) \times \ldots \times (n - r + 1)$ is the same as $\dfrac{n!}{(n - r)!}$. Let's look at this by example first, to try and convince you that the two things are the same.

Consider ordering 3 elements from a list of 10 elements, as above. We know the answer is $10 \times 9 \times 8$. In this example $n = 10$ and $r = 3$. The formula $\dfrac{n!}{(n - r)!}$ is therefore

$$\frac{10!}{(10 - 3)!} = \frac{10!}{7!} = \frac{10 \times 9 \times 8 \times 7 \times 6 \times 5 \times 4 \times 3 \times 2 \times 1}{7 \times 6 \times 5 \times 4 \times 3 \times 2 \times 1}$$

Now, we can cancel $7 \times 6 \times 5 \times 4 \times 3 \times 2 \times 1$ from the top and bottom, and we are just left with $10 \times 9 \times 8$ as before, so this is the same thing.

Take the other examples and try to convince yourself that these two things are the same. We usually write permutations in the second way as $\dfrac{n!}{(n-r)!}$, as it's easier to write and easier to calculate.

You will often see the notation nPr for permutations: the n and r are the variables and P stands for permutations. To make the n and r distinctive, they are often written above and below the P as ^nP_r, which is the notation we shall use in this book. You can also write this in other ways. Get used to this; people do use different notations.

So, for example, $^{10}P_3$ is the number of permutations of 3 elements from 10 elements, which is given by $\dfrac{10!}{(10-3)!} = 720$, just as we did above.

If you have a scientific calculator then you will probably have a ^nP_r button. Can you find it and work out how to use it?

Learn to use your calculator. If you have a scientific calculator that can do such calculations, find the button and practise with it until you understand how it works. Read the instructions, search for help online, ask a friend – but learn how to use it correctly to save you time and effort.

To summarise this, the following is the basic definition you need:

The number of permutations of r elements from a set of n elements is $^nP_r = \dfrac{n!}{(n-r)!}$

Note that, if we take $n = r$, then we are asking the question as to how many ways there are to order n elements from a set of n elements or, simply, how many possible orders of all the elements there are. We worked out in the factorials section that this is just $n!$. But using this ^nP_r formula, we get $\dfrac{n!}{(n-n)!} = \dfrac{n!}{0!}$. Remembering that we defined $0!$ to be 1, then this is just $\dfrac{n!}{1} = n!$, which matches the answer we gave before. So it makes sense to define $0! = 1$, as this definition still works nicely, and fits with everything else we are doing. There is a reason for everything!

● Combinations

In the previous section we concerned ourselves with permutations – that is, where the order matters. When considering first, second and third places, an ordering of A, D, E is different from an ordering of D, E, A for example.

However, what happens if we don't care about the order, but simply want to select a certain number of elements, no matter what order they are in? For example, suppose we want to choose 3 out of 5 people to be selected for the next round of the competition, without ranking them first, second or third, so just selecting 3 from the 5 people. How many possible ways are there to do this?

The crucial difference here is that things such as A, D, E and D, E, A are now the same: these three people (A, D and E) progress, and the others (B, C) miss out.

So this is like a permutation question, except that we no longer care about order. How many things are 'the same' here? If we are choosing 3 people, then there are $3! = 6$ permutations of the same set of people that are just considered to be the same here.

So to work out the number of possibilities, we could first of all work out all the possibilities including order, which is ^nP_r, and then divide this number by 6, since every possibility (e.g. ADE) appears 6 times in the list of all permutations (ADE, AED, DAE, DEA, EAD, EDA in this example).

To illustrate, in our example, the number of ways is $\dfrac{5 \times 4 \times 3}{6} = 10$, and so there are 10 ways to choose 3 people, not caring about the order (in actual fact these are ABC, ABD, ABE, ACD, ACE, ADE, BCD, BCE, BDE, CDE; there are no other possibilities if order does not matter).

Generalising this, we can calculate the number of choices (not caring about order) to be $\dfrac{^nP_r}{r!}$.

We call a choice where we are not concerned about the order a *combination*, and use the notation nCr or ^nC_r to stand for this. If you remember that $^nP_r = \dfrac{n!}{(n-r)!}$, then this means we are defining ^nC_r as this divided by $r!$, or $\dfrac{n!}{r!(n-r)!}$. So:

> The number of combinations of r elements from a set of n elements is $^nC_r = \dfrac{n!}{r!(n-r)!}$

Note, that if $r = n$ then we are asking how many ways there are to pick n elements from n elements where order doesn't matter. It's fairly obvious that the answer is 1. If order doesn't matter, then there is only one way to pick n elements – that is to pick them all. If you put this into the formula, you will also get the answer 1. You get:

$$^nC_n = \frac{n!}{n!(n-n)!} = \frac{n!}{n!0!} = \frac{n!}{n! \times 1} = \frac{n!}{n!} = 1,$$

so the definition does make sense here as well.

> Remember: the important point is that with permutations we care about the order, but with combinations the order doesn't matter.

Let's look at some examples.

Example

There are 8 competitors in an Olympic final. How many ways are there to allocate the gold, silver and bronze medals?

Solution

In this question, the order matters, and so we need permutations. The number of ways to allocate the medals is $^8P_3 = \dfrac{8!}{(8-3)!} = \dfrac{8!}{5!} = 336$.

Now consider a situation where order doesn't matter.

Example

There are 8 competitors in an Olympic qualifying event. How many ways are there to determine 3 competitors to proceed to the next round?

Solution

In this question, the order does not matter, and so we need combinations. The number of ways to determine the qualifiers is $^8C_3 = \dfrac{8!}{3!(8-3)!} = \dfrac{8!}{3!5!} = 56$.

Note that there are always fewer combinations than permutations. This makes sense, since with combinations several permutations are being considered as the same thing, and so there are clearly going to be fewer combinations.

● Further notes on combinations

Note that $^8C_3 = \dfrac{8!}{3!(8-3)!} = \dfrac{8!}{3!5!} = 56$, and also that $^8C_5 = \dfrac{8!}{5!(8-5)!}$ $= \dfrac{8!}{5!3!} = 56$, and the two answers are the same. This means that the number of ways of choosing 3 elements from 8 elements is the same as the number of ways of choosing 5 elements from 8 elements.

Is this coincidence? No, it makes perfect sense. Choosing the 3 competitors to proceed is really just the same as choosing the 5 competitors to miss out (instead of declaring the winners, you declare the ones eliminated), and so it makes logical sense that the answers should be the same.

In general, if you work out ^nC_r and $^nC_{n-r}$ you will always get the same answer.

Note also that, for any n, it will always be the case that $^nC_0 = 1$ and $^nC_n = 1$: there is only one way to choose no objects (just don't choose anything) and only one way to choose n objects (choose them all).

Combinations are heavily used in mathematics, and the alternative notation $\binom{n}{r}$ is often used instead of ^nC_r or nCr, although you are less likely to see this in more statistical subjects. But make sure you know what notation is being used.

✔ There isn't always one standard notation for a topic; lecturers and authors may use different notation for the same concepts. This can be annoying, but it is a fact of life. It isn't anything to worry about as long as you pay attention, be aware of the notation they are using, and understand that the concept is exactly the same, regardless of the precise notation being used.

Permutations and combinations feature heavily in statistics, and also in many other branches of mathematics. Remember that permutations insist on an order, whereas combinations disregard order. If you then just learn the formulae you can use them confidently.

> To be able to use these concepts you really need just to learn the definitions of factorial and the formulae for ^nPr and ^nCr. But I've tried to give you the idea of where these formulae came from, and why they are relevant. If you understand (at least informally) where a formula comes from, and why we use the definitions we do, then the subject is much more natural and easier to learn.

 Exercises

You can use a calculator for these exercises.

1 What are the following factorials?

Example: 5!

Solution: $5! = 5 \times 4 \times 3 \times 2 \times 1 = 120$

(a) 4!   (b) 6!   (c) 10!   (d) 2!   (e) 1!   (f) 0!

2 Calculate the following:

Example: $^7P_5$ and $^7C_5$

Solution: $^7P_5 = \dfrac{7!}{(7-5)!} = 2520$ and $^7C_5 = \dfrac{7!}{5!(7-5)!} = \dfrac{7!}{5!2!} = 21$

(a) $^6P_3$ and $^6C_3$ (b) $^8P_5$ and $^8C_5$ (c) $^{10}P_4$ and $^{10}C_4$

(d) $^5P_2$ and $^5C_2$ (e) $^8P_1$ and $^8C_1$ (f) $^5P_5$ and $^5C_5$

3 Answer the following questions using permutations.

Example: In a swimming final there are 9 swimmers. In how many ways can the gold, silver and bronze medals be awarded?

Solution: The order matters here, so we need permutations. We need to work out $^9P_3 = \dfrac{9!}{(9-3)!} = 504$, and so there are 504 possible ways.

(a) There are 8 children in a group. In how many ways can two of these be selected for a winner and a runner-up prize?

(b) There are 6 power units with cables, and 4 sockets placed in order. In how many ways can the units be connected to the sockets?

(c) A 2010 Formula 1 Grand Prix race has 24 cars. In how many ways can the drivers fill the top 10 point-scoring positions?

4 Answer the following questions using combinations.

Example: In a 'talent' show there are 10 competitors. In how many ways can 4 of these be chosen to proceed to the final?

Solution: The order does not matter here, so we need combinations. We need to work out $^{10}C_4 = \dfrac{10!}{4!(10-4)!} = \dfrac{10!}{4!6!} = 210$, and so there are 210 possible ways.

(a) There are 7 medicines available. In how many ways can 3 of these be selected for a scientific test?

(b) How many ways are there to select 5 out of 10 clients for a bonus voucher?

(c) Out of 18 lenders, it is necessary to choose 5 of them for a survey. In how many ways can this be done?

5 *(Harder) Additional – you will have to decide if you need to use factorials, permutations, combinations, or a mixture of approaches.*

(a) A password is made up of 5 *different* letters from the English alphabet. How many different passwords are there?

(b) How many ways are there to place 8 books in order on a bookshelf?

(c) In how many ways can you choose 5 employees from your staff of 9 people to take part in a training exercise?

(d) How many different possibilities of lottery numbers are there? (In the lottery, 6 balls are drawn from 49 balls; order does not matter.)

(e) The 2010 England World Cup squad contained 3 goalkeepers, 8 defenders, 8 midfielders and 4 strikers. In how many ways could the manager pick the team of 11, assuming that he always chose the team in the same formation, which consisted of 1 goalkeeper, 4 defenders, 4 midfielders and 2 strikers?

Sigma notation

The use of 'sigma notation' to simplify mathematical expressions and present them concisely

We have extensively discussed measurements that involve 'adding a lot of things up'. Here we shall formalise this and introduce sigma notation, which we shall use in the rest of the book (although always accompanied with an informal explanation) to give concise mathematical formulae for the concepts we need.

Key topics
- Series and sequences
- Sigma notation
- Expressing formulae in sigma notation

Key terms
sequence series sigma notation pi notation

In what we have done so far, we have often used phrases such as 'add the values together' and similar slightly vague instructions. Here we are going to formalise this, using what is known as *sigma notation*. It is imperative that you get to grips with this, because your statistics material is very likely to use it. I'll try to give you a gentle introduction to it, so that you have no reason to be worried when you do come across it.

● Sequences

A *sequence* is a list of numbers (often called *terms*, although names such as *elements* are often used as well), usually with some pattern to it. Sequences can be either finite (a finite number of terms) or infinite.

For example, the sequences 1, 2, 3, 4, 5 and 1, 4, 9, 16, 25, 36 are finite sequences: they contain 5 and 6 terms respectively.

Infinite sequences are often denoted by writing down the first few terms (enough to show what the pattern is) and then using dots to indicate that the sequence continues for ever. Examples of infinite sequences are given below. Can you work out what is the next term in each case?

- 2, 4, 6, 8, 10, …
- 1, 4, 7, 10, 13, …
- 1, 2, 4, 8, 16, …

You can give a sequence a name. Let us say, for example, that s is the sequence 2, 4, 6, 8, 10, … . You can refer to particular terms of the sequence using subscripts. So s_1 is the first term of the sequence, which is 2; s_2 is the second term of the sequence, which is 4; similarly $s_3 = 6$, $s_4 = 8$, and so on. Note that in this example the pattern is simple, and you can write down a general formula for a particular term: the ith term is $2i$ (e.g. the 4th term is $2 \times 4 = 8$), and so you can write a definition of s as the sequence where $s_i = 2i$. (Note that i is often used as a subscript letter, but you can use any other letter if you like.)

● Series

A *series* is similar to a sequence, except that instead of listing all the terms, we add them up.

Similarly to sequences, examples of finite series are $1 + 2 + 3 + 4 + 5$ (which makes 15) and $1 + 4 + 9 + 16 + 25 + 36$ (which makes 91). So, with series, we have a definite 'answer' to the sum.

You might think that infinite series are rather pointless. The sum of $1 + 2 + 3 + 4 + \dots$ is obviously going to be infinite; surely, if we add up infinitely many things, then the answer will be infinite?

Actually this is not true; there are infinite series that *do* have a finite answer. This wasn't understood until the 17th and 18th centuries with the development of calculus and the idea of limits. A good example is the series $1 + \frac{1}{2} + \frac{1}{4} + \frac{1}{8} + \frac{1}{16} + \dots$, in which each term is half of the previous one.

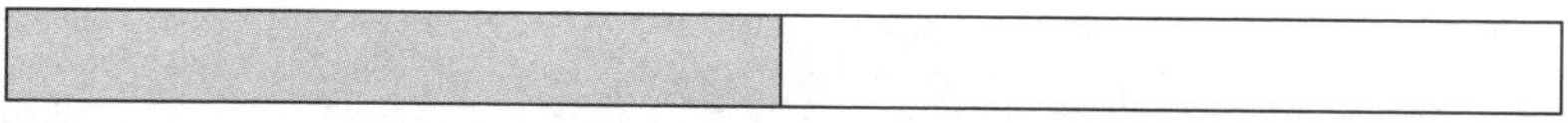

Figure 8.1 Visualisation of a series showing 1 term removed

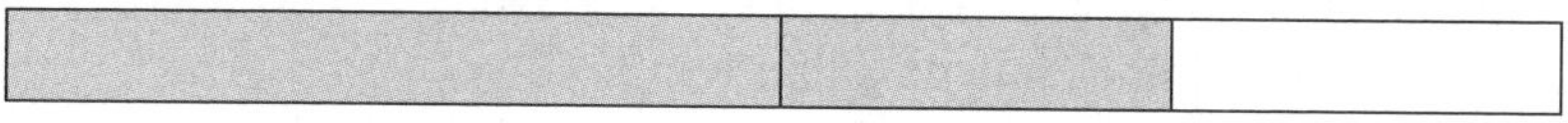

Figure 8.2 Visualisation of a series showing 2 terms removed

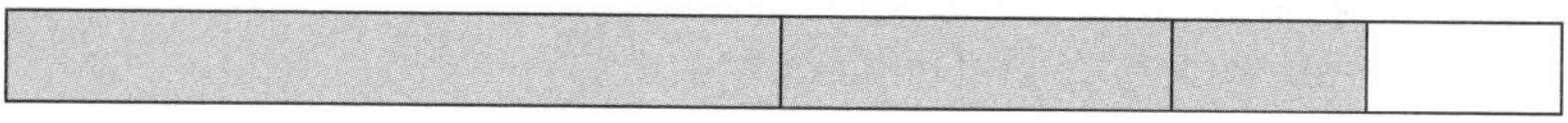

Figure 8.3 Visualisation of a series showing 3 terms removed

A nice way to think what the answer to this infinite sum should be is to visualise a piece of wood 2 m long, and imagine that the terms in the series are pieces of wood that you cut away. So first of all you cut 1 m away (Figure 8.1). Next you cut away $\frac{1}{2}$ m from what is left (Figure 8.2), and then $\frac{1}{4}$ m from what is left (Figure 8.3), and so on, each time cutting away half of what is left. You can see that you can keep cutting for ever, always cutting half of what is left, and so you'll never get past 2 m. Therefore it makes sense that the answer to $1 + \frac{1}{2} + \frac{1}{4} + \frac{1}{8} + \frac{1}{16} + \dots$ is 2.

This is an example of what is known as the *sum to infinity of a geometric series*. We shall not discuss this topic in this book, but if you are interested, do some research and find out more about it.

Interestingly, the series $1 + \frac{1}{2} + \frac{1}{3} + \frac{1}{4} + \frac{1}{5} + \dots$ does give an infinite answer, unlike the series above.

Rather than learning by rote just those things you need for a course or exam, try to immerse yourself in a subject. This mathematics has a history, and someone developed the ideas. Find out about the history. If you find something interesting, get another book, or go online and find out something more. As well as providing interest, it helps you understand a topic and its relevance much more than simply learning something to pass an exam.

● Sigma notation

Series appear a lot in statistics, but this leaves us with notational difficulties. If we have several of them, it gets long and awkward to keep writing values and dots like $1 + 2 + 3 + 4 + 5 + \ldots$ everywhere, especially when the formulae get long. *Sigma notation* is a concise mathematical way to write series. It takes a bit of practice to get your head around it fully but, once mastered, it makes things much easier to denote and explain.

A sigma is a capital Greek letter, written as Σ. Greek letters appear a lot in mathematics and statistics (a lot of mathematics was developed in ancient Greece, and we run out of letters if we just use English ones), so it is worth knowing the symbols. A list of Greek symbols is given in the appendices.

How can we use sigma notation to write down series? Let's start with simple finite series. The series $1 + 2 + 3 + 4 + 5$ has five terms. If we call this series s, then we can use subscript notation as before. The ith term is simply i (so $s_1 = 1$, $s_2 = 2$, and so on).

The general form of sigma notation is to write $\sum_{i=1}^{n} s_i$ where n is the number of elements in the series; so in this case $n = 5$. What this notation means is to write down what is after the sigma, start with $i = 1$: so write down s_1. Then increase i by 1, so i becomes 2, and then again write down what is after the sigma, so write down s_2 (i is now 2, so we write s_2), and so on. We keep writing them down, so we would write down s_3, s_4 and s_5, until we reach the ending point n (which is 5 in this case). Then we add up these values.

So, to sum up, $\sum_{i=1}^{5} s_i = s_1 + s_2 + s_3 + s_4 + s_5$, which is exactly the definition of the series. In this case there is a simple formula for the s_i, namely that $s_i = i$. So we can express the series $1 + 2 + 3 + 4 + 5$ in sigma notation as $\sum_{i=1}^{5} i$.

I'll now do some more examples, to help you get comfortable with the idea. You should focus on being able to understand what the sigma notation means. It's hard to 'create' the sigma notation from a given series, but at this stage, provided you understand what an expression in sigma notation is telling you, then that should be enough for you to understand texts and examples, without you having to write things in sigma notation yourself.

Example 1

What is the series $\sum_{i=1}^{6} 2i$?

Solution

Start with $i = 1$, and write down what is after the sigma. This is $2i$, which is 2×1 (since $i = 1$), so we write down 2. Next, increase i by 1, and so $i = 2$. Write down what is after the sigma: this is $2i$, which is 2×2 (since $i = 2$), so we write down 4. Continuing, we write down 6 (when $i = 3$), 8 (when $i = 4$), 10 (when $i = 5$) and 12 (when $i = 6$), which is where we stop, as $i = 6$ is the stopping point, as indicated on top of the sigma. Adding them all up, this means we have written down the series $2 + 4 + 6 + 8 + 10 + 12$ (which adds up to 42).

Example 2

What is the series $\sum_{i=1}^{10} i^2$?

Solution

Start with $i = 1$, and write down what is after the sigma, giving i^2, which is 1^2 (since $i = 1$), so we write down 1. Next, increase i by 1 and so $i = 2$. Write down what is after the sigma, giving i^2, which is 2^2 (since $i = 2$), so we write down 4. Continuing, we write down 9 (when $i = 3$), 16 (when $i = 4$), and so on until we reach the stopping point of $i = 10$, as indicated on top of the sigma, so the last term we write down is $10^2 = 100$. Adding them all up, this means we have written down the series $1 + 4 + 9 + 16 + 25 + 36 + 49 + 64 + 81 + 100$ (which adds up to 385).

Hopefully, you are starting to get the idea; for long series of data, it is much easier to write them in sigma notation than to write them out in full.

To denote infinite series, you need to use the infinity symbol ∞.

Example 3

What is the series $\sum_{i=1}^{\infty} \frac{1}{2^i}$?

Solution

Start with $i = 1$, and write down what is after the sigma, $\frac{1}{2^i}$, which is $\frac{1}{2^1}$ (since $i = 1$), so we write down $\frac{1}{2}$. Next, increase i by 1 and so $i = 2$. Write down what is after the sigma, so $\frac{1}{2^i}$, which is $\frac{1}{2^2}$ (since

$i = 2$), so we write down $\frac{1}{4}$. We have to keep going like this for ever (the stopping point is infinity, so we just keep adding for ever), so we end up with the series $\frac{1}{2} + \frac{1}{4} + \frac{1}{8} + \frac{1}{16} + \ldots$ (which adds up to 1, using a similar argument to the example we did before with cutting the pieces of wood).

How would you express the series $1 + \frac{1}{2} + \frac{1}{4} + \frac{1}{8} + \frac{1}{16} + \ldots$ (which we saw before) in sigma notation? There are various ways to do it, but the two most obvious are below:

- Note that the ith term of this sequence is $\frac{1}{2^{i-1}}$: so the first term is $\frac{1}{2^{1-1}} = \frac{1}{2^0} = \frac{1}{1} = 1$ (since anything to the power 0 is always 1), and so on for the other terms. So we can use sigma notation to write this as $\sum\limits_{i=1}^{\infty} \frac{1}{2^{i-1}}$. Write this out for yourself to convince yourself that it works.

- Alternatively, you can change the starting point and start instead with $i = 0$. If you write $\sum\limits_{i=0}^{\infty} \frac{1}{2^i}$ then the first term is for $i = 0$, and so you get $\frac{1}{2^0} = \frac{1}{1} = 1$ as the first term, and so on.

Which of these approaches you take is up to you; they are the same thing. In complex problems you would try to keep things consistent to help simplify things, so that everything starts from the same thing, be it $i = 0$ or $i = 1$. For now you just need to understand that there are various ways to express things, and that you should express them in whichever way you find convenient.

✔ Often in mathematics and statistics, there is more than one way to express things. Don't despair if you have written your answer in a different form from the answer in the back of a book, as long as it is correct. Just as writing $y + x$ rather than $x + y$ in an algebraic expression is not wrong at all, nor is writing your sigma notation in a slightly different form. As long as what you write is correct and you understand it, it is fine – unless the question specifically asks for a particular form, say starting from $i = 1$, but you don't get too many questions like this.

We can express some of the statistical concepts we have seen so far in sigma notation.

● The mean in sigma notation

Recall that the mean of a list of data is worked by adding up all the values, and then dividing by the number of values.

Suppose we have a list of data x with terms $x_1, x_2, x_3, \ldots, x_n$, and so there are n terms (n pieces of data). Then we want to add up all these x_i and then divide by n (the number of terms) and so we can give the formula for the mean in sigma notation as follows:

The mean of the list of data $x = x_1, x_2, x_3, \ldots, x_n$ is $\dfrac{\sum\limits_{i=1}^{n} x_i}{n}$, or it can be written as $\dfrac{1}{n} \sum\limits_{i=1}^{n} x_i$, which is perhaps slightly easier to write without using up too much vertical space.

Look at this carefully and convince yourself that all it is saying is 'add up the values and then divide by the number of values', which is the definition of the mean. We are going to use sigma notation a lot, so make sure you understand it and how it works, so that you can be confident looking at formulae that use it.

It is common to denote the mean of a list of data x as $\bar{x}$: that is, a bar above the x. Remember this notation, as we shall use it again.

● The variance in sigma notation

Note: In everything we do here, we are considering the variance and standard deviation to be the population variance and population standard deviation. The sample variance and sample standard deviation are almost exactly the same, but with $n - 1$ instead of n.

Recall that the variance is worked out by taking the difference between every value and the mean, and squaring them, and then adding them all up and dividing by the number of values. Using the notation $\bar{x}$ for the mean as discussed above, we can create a formula for the variance.

$$\text{The variance of the list of data } x = x_1, x_2, x_3, \ldots, x_n \text{ is } \dfrac{\sum_{i=1}^{n}(x_i - \bar{x})^2}{n}, \text{ or it}$$

can be written as $\dfrac{1}{n}\sum_{i=1}^{n}(x_i - \bar{x})^2$, which is again perhaps slightly easier to write without using up too much vertical space.

Look carefully at this formula and try to realise what it is saying. For each value, take the difference from the mean, and square the answer. Then add up all the values, and divide the total by n. This is exactly the definition of the variance, but formalised in mathematical notation.

● An alternative formula for the variance

It is possible to show that this formula for the variance is the same

as the formula $\dfrac{\sum_{i=1}^{n}x_i^2}{n} - \bar{x}^2$ (or $(\dfrac{1}{n}\sum_{i=1}^{n}x_i^2) - \bar{x}^2$), which gives us a faster

way to work out the variance: we add up all the squares of the values and divide by n, and then subtract the mean squared. The advantage of this formulation is that you don't have to take the mean away at every step; just take away its square at the end.

I won't prove here why these are equivalent; but if you are confident in your mathematical ability, then feel free to have a go. But I will illustrate by example:

Example

Work out the variance of the set of data 4, 5, 9, 10.

Solution 1

Using the first formulation of the variance $\dfrac{\sum_{i=1}^{n}(x_i - \bar{x})^2}{n}$, we need to take the difference of every value from the mean, and then divide by n, the number of values, which is 4 in this case. The mean is $\dfrac{4 + 5 + 9 + 10}{4} = \dfrac{28}{4} = 7$. Hence we need to work out the squared differences and divide by 4 (the number of values), and so work out $\dfrac{(4 - 7)^2 + (5 - 7)^2 + (9 - 7)^2 + (10 - 7)^2}{4} = \dfrac{(-3)^2 + (-2)^2 + 2^2 + 3^2}{4} =$

$$\frac{9 + 4 + 4 + 9}{4} = \frac{26}{4}$$

which is 6.5 as a decimal, and so the variance is 6.5.

Solution 2

The second formulation of the variance is $\frac{\sum_{i=1}^{n} x_i^2}{n} - \bar{x}^2$, so we add up all the squares of the values, divide by n, and then take away the mean squared. So, in this example (remember that the mean is 7), we work out $\frac{4^2 + 5^2 + 9^2 + 10^2}{4} - 7^2$, which is $\frac{16 + 25 + 81 + 100}{4} - 49$

$= \frac{222}{4} - 49 = 55.5 - 49 = 6.5$

and we get the same answer for the variance.

You can use either formulation. The second is probably a little easier to calculate, as you don't have to take the mean away at every step, but the first is probably more 'natural' in terms of what variance *means* (the distance of values away from the average).

● The standard deviation in sigma notation

Remembering that the standard deviation is simply the square root of the variance, then the formula for the standard deviation is just

$$\sqrt{\frac{\sum_{i=1}^{n}(x_i - \bar{x})^2}{n}} \quad \text{or} \quad \sqrt{\frac{\sum_{i=1}^{n} x_i^2}{n} - \bar{x}^2}$$

depending on which version of the variance formula you prefer to use.

● Pi notation

There is a similar notation to sigma notation that is used when you multiply things together, rather than add them up. This notation uses a capital 'pi' symbol (again, a Greek letter), and so for example $\prod_{i=1}^{5} i$ means $1 \times 2 \times 3 \times 4 \times 5$ (which is 120). You probably won't encounter this again (it's nothing like as common as sigma notation), but I'm mentioning just in case you do come across it and need to know what it means.

Summary

Sigma notation takes some getting used to, but it is a very concise, effective way of representing series, which crop up all the time in statistics. Practise with it and become comfortable with what something in sigma notation means, and you will see that you can write quite complex formulae in a concise mathematical way, which is very important when it comes to bigger problems.

Exercises

1 Write down the next term in each of the following sequences:

Example: 2, 5, 8, 11, 14, ...

Solution: In this sequence, you can spot that the difference between each term is 3, so the next term is 17.

(a) 1, 5, 9, 13, 17, ...

(b) 1, 4, 9, 16, 25, ...

(c) 1, 10, 100, 1000, ...

(d) $\frac{1}{2}, \frac{1}{3}, \frac{1}{4}, \ldots$

(e) 1, 3, 9, 27, ...

(f) 1, 1, 2, 3, 5, 8, 13, 21, ...

(*Hint for (f): this is known as the Fibonacci sequence, so research this if you can't spot the pattern.*)

2 Write out the following series (given in sigma notation) in full. Can you actually perform the addition and give an answer?

Example: $\sum\limits_{i=1}^{5} 3i$

Solution: First i is 1, so we write down the value of the expression $3i$, which is $3 \times 1 = 3$. Then we consider $i = 2$, so we write down 3×2 (since $i = 2$), which is 6. Similarly, when $i = 3$, we write down 9; when $i = 4$, we write down 12; and when $i = 5$ (the stopping point) we write down 15. Hence this series is $3 + 6 + 9 + 12 + 15$, which you can calculate to be 45.

(a) $\sum\limits_{i=1}^{6} 4i$

(b) $\sum\limits_{i=1}^{4} (2i + 1)$

(c) $\sum\limits_{i=0}^{4} 2^i$

(d) $\sum\limits_{i=5}^{10} (i + 1)^2$

(e) $\sum\limits_{i=0}^{\infty} \frac{1}{2^{i+1}}$

(f) $\sum\limits_{i=1}^{\infty} i$

3 Use both formulae for the variance to calculate the variance and standard deviation of the following lists of data.

Example: 1, 3, 6, 7, 8

Solution: The mean is $\frac{1 + 3 + 6 + 7 + 8}{5} = \frac{25}{5} = 5$. Using the first formula, we work out the difference between every value and the mean and square them, before dividing by the number of values (5), so we work out the variance to be

$$\frac{(1 - 5)^2 + (3 - 5)^2 + (6 - 5)^2 + (7 - 5)^2 + (8 - 5)^2}{5}$$

$$= \frac{16 + 4 + 1 + 4 + 9}{5} = \frac{34}{5} = 6.8.$$

Using the second formula, we square all of the values, divide by the number of values (5), and subtract the mean squared, and so we work out the variance to be

$$\frac{1^2 + 3^2 + 6^2 + 7^2 + 8^2}{5} - 5^2 = \frac{1 + 9 + 36 + 49 + 64}{5} - 25 =$$

$$\frac{159}{5} - 25 = 31.8 - 25 = 6.8.$$

Either way, we work out the variance to be 6.8, and so the standard deviation is $\sqrt{6.8} = 2.608$ (to 3 decimal places)

(a) 1, 4, 7, 9, 10, 13   (b) 2, 3, 8, 9, 10

(c) 2, 5, 5, 5, 6, 8, 9, 10   (d) 4, 4, 4, 4, 4, 4

4 Write the following series using sigma notation:

Example: 3 + 6 + 9 + 12 + 15.

Solution: The ith term is $3i$ (the first term is 3×1, the second term is 3×2, etc.) and so, since there are five terms, this is $\sum_{i=1}^{5} 3i$.

(a) 4 + 8 + 12 + 16 + 20 + 24 + 28   (b) 1 + 2 + 3 + 4 + 5 + ...

(c) 1 + 3 + 5 + 7 + 9 + 11   (d) $2 + 1 + \frac{1}{2} + \frac{1}{4} + \frac{1}{8} + ...$

CORRELATION AND REGRESSION

Giving a measure of relationship between two variables

Is there a connection between two variables? Here we shall give formulae to calculate a measure of how closely two variables are related.

Key topics
● Scatter diagrams
● Correlation coefficients
● Ranking

Key terms
scatter diagram correlation positive and negative correlation
Pearson's correlation coefficient Spearman's rank correlation
coefficient

In Chapter 2 we discussed scatter diagrams, and briefly mentioned the idea of correlation. Here we shall formalise the idea of correlation, by giving clear formulae to measure the relationship between two variables.

● Revision of scatter diagrams and correlation

Recall that in Chapter 2 we discussed a sample of people of varying heights, and we noticed that there seemed to be a *correlation* (a relationship) between height and weight, but not between height and the number of shirts owned. This makes perfect sense; taller people are likely to be heavier, but why would a person's height affect the number of shirts they own?

Here is some more data, which gives the mark obtained by a sample of 10 students in an examination at the end of a course. The number of hours of revision they did, the month of the year they were born (from 1 to 12, January to December), and the number of classes they missed through the year are also given.

Revision hours	18	2	13	14	6	15	16	9	10	15
Month of birth	1	9	10	3	2	11	4	6	7	12
Classes missed	3	30	20	7	24	1	5	16	10	12
Mark	82	20	42	68	41	95	72	48	60	62

We shall plot three scattergraphs (as in Chapter 2), which show the mark obtained plotted against:

(a) hours of revision (Figure 9.1);

(b) month of birth (Figure 9.2);

(c) number of classes missed (Figure 9.3).

Just by looking at these graphs, it's pretty clear that:

- In (a), there seems to be a positive correlation between the number of hours of revision and the mark obtained; so the students that revised more, did better.

- In (b), there seems to be no correlation; the month of birth makes no difference.

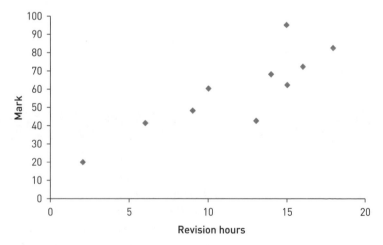

Figure 9.1 **Example of a scattergraph showing hours of revision**

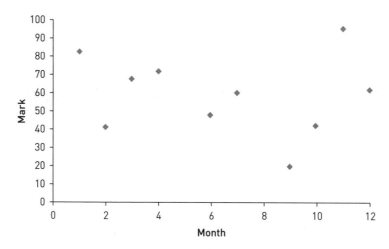

Figure 9.2 **Example of a scattergraph showing month of birth**

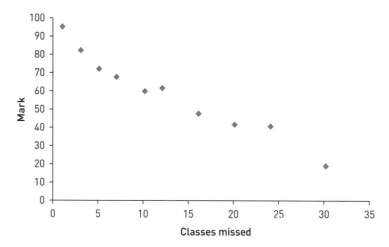

Figure 9.3 **Example of a scattergraph showing the number of classes missed**

- In (c), there seems to be a (quite strong) negative correlation between classes missed and the marks: the more classes students missed, the worse they did.

Look at these graphs and the conclusions. Do they all make sense? Yes, they do. So in your studies, don't worry about irrelevant things like when you were born, your height, weight, or your age; all that matters is you, the work you put in, and your effort to attend classes and learn.

However, we are making these judgements only by looking at a graph. Let's now look at how to formalise this mathematically, and give a proper measure of how closely correlated two variables are.

● Correlation coefficients

We'd like to create a numerical way to measure the correlation of a sample of data such as the one above. Ideally, our measure of correlation will take a value between -1 and 1, where 1 represents perfect positive correlation, -1 represents perfect negative correlation, and values in between represent varying levels of correlation (including 0, where there is no correlation at all), so that a value such as 0.9 represents a very strong positive correlation, a value such as -0.5 represents a weak negative correlation, and so on.

The most common measure of correlation is the *Pearson correlation coefficient* (formally, *Pearson's product moment correlation coefficient*, but people will know what you mean if you refer just to the Pearson correlation coefficient).

I shall illustrate how it works by using the example of revision hours and marks as before; recall that we saw from the scattergraph that we expect a positive correlation.

Revision hours	18	2	13	14	6	15	16	9	10	15
Mark	82	20	42	68	41	95	72	48	60	62

We shall call the revision hours x_i, so that x_1 is the first value (18), x_2 is the second value (2), and so on. Similarly, we shall call the mark y_i, so that y_1 is the first value (82), y_2 is the second value (20), and so on.

First we need to work out the means of each of these. Using the usual formula for the mean, you can calculate the mean of the revision hours to be 11.8, and the mean of the marks to be 59 (check that you can get these values). We shall denote these by $\bar{x}$ and $\bar{y}$ respectively, so $\bar{x} = 11.8$ and $\bar{y} = 59$.

Now we are going to work out a sum that is given in sigma notation as $\sum_{i=1}^{n}(x_i - \bar{x})(y_i - \bar{y})$. What this means is that, for each value x_i, calculate its difference from the mean $\bar{x}$, and similarly for each

y_i, calculate its difference from the mean $\bar{y}$, then multiply the two resulting numbers together. Finally add all these products up. Note that this is very similar to the calculation of the variance, but we have two data sets here.

We can illustrate this on the next few rows of our table:

Revision hours (x_i)	18	2	13	14	6	15	16	9	10	15	
Mark (y_i)	82	20	42	68	41	95	72	48	60	62	Sum
$x_i - \bar{x}$	6.2	−9.8	1.2	2.2	−5.8	3.2	4.2	−2.8	−1.8	3.2	0
$y_i - \bar{y}$	23	−39	−17	9	−18	36	13	−11	1	3	0
$(x_i - \bar{x})(y_i - \bar{y})$	142.6	382.2	−20.4	19.8	104.4	115.2	54.6	30.8	−1.8	9.6	837

In the first column, we work out $x_1 - \bar{x} = 18 - 11.8 = 6.2$, and $y_1 - \bar{y} = 82 - 59 = 23$, and then we multiply these numbers together to get $(x_i - \bar{x})(y_i - \bar{y}) = 6.2 \times 23 = 142.6$. We do the same for the rest of the columns. After doing all these, add them all up; you should get the sum to be 837.

Note that I have included the sums of the $x_i - \bar{x}$ and the $y_i - \bar{y}$, which work out to be 0. They should always be zero (by the definition of the mean – can you see why?), so this acts as a useful check that we have got it right.

Next we need to work out something similar to the variance of the x_i and y_i by adding up all the 'differences squared'. The only difference with the variance is we aren't dividing by n, so we are going to work out $\sum_{i=1}^{n}(x_i - \bar{x})^2$ and $\sum_{i=1}^{n}(y_i - \bar{y})^2$.

We can do this as another two rows to our table. All we are really doing is squaring the values in the third row and adding them up, and the same for the fourth row.

Revision hours (x_i)	18	2	13	14	6	15	16	9	10	15	
Mark (y_i)	82	20	42	68	41	95	72	48	60	62	Sum
$x_i - \bar{x}$	6.2	−9.8	1.2	2.2	−5.8	3.2	4.2	−2.8	−1.8	3.2	0
$y_i - \bar{y}$	23	−39	−17	9	−18	36	13	−11	1	3	0
$(x_i - \bar{x})(y_i - \bar{y})$	142.6	382.2	−20.4	19.8	104.4	115.2	54.6	30.8	−1.8	9.6	837
$(x_i - \bar{x})^2$	38.44	96.04	1.44	4.84	33.64	10.24	17.64	7.84	3.24	10.24	223.6
$(y_i - \bar{y})^2$	529	1521	289	81	324	1296	169	121	1	9	4340

Finally, we can work out the Pearson correlation coefficient. The formula, which just uses the three sums we have worked out, is

The Pearson correlation coefficient

$$r = \frac{\sum_{i=1}^{n}(x_i - \bar{x})(y_i - \bar{y})}{\sqrt{\sum_{i=1}^{n}(x_i - \bar{x})^2}\ \sqrt{\sum_{i=1}^{n}(y_i - \bar{y})^2}}$$

You can easily work this out. What's on the top is the sum 837, and the square roots on the bottom of the formula are the square roots of the sums 223.6 and 4340: these are the three values in our sum column.

So we need to work out $\dfrac{837}{\sqrt{223.6}\ \sqrt{4340}} = 0.850$ (to 3 decimal places).

0.850 is a high value – quite close to 1 – and would seem to definitely show evidence of a strong positive correlation between the amount of revision and the mark obtained. You could do a formal test on this value, but there is no need for us to discuss this at this stage.

This took a long time, but that's because I was talking you through it. In practice, you can essentially answer the question just by making the table.

Now try to work out the correlations for the other two data sets (month of birth, and classes missed). It might take a while for you to draw the table, but it will make you understand how we are doing it, and why we are doing it.

If you get it right, you should get −0.106 for the month of birth, which is close to zero, and so there is virtually no correlation, and −0.974 for classes missed, which is extremely close to −1, and so there is a very strong negative correlation.

There are software commands that can do all this for you. But this book is about making you understand the ideas behind the topics. I want you to do these questions, and get them right, so that you understand what something means, rather than blindly press a button on a computer without having any idea of what you are actually doing.

● Alternative formulation

Recall that, back in Chapter 5, we gave alternative formulations for the variance. There is also an alternative variation for Pearson's correlation coefficient, which is probably a little easier to work out as it doesn't involve so many differences, but is a lot less intuitive in terms of what it means, and probably looks more complicated. I'll give you the formula before we break it down as before:

The Pearson correlation coefficient can also be given by the formula

$$r = \frac{\left(\sum_{i=1}^{n} x_i\, y_i\right) - n.\bar{x}.\bar{y}}{\sqrt{\left(\sum_{i=1}^{n} x_i^2\right) - n\bar{x}^2}\sqrt{\left(\sum_{i=1}^{n} y_i^2\right) - n\bar{y}^2}}$$

Note, by the way, that you can combine the square roots in the bottom of this fraction if you wish, and write

$$r = \frac{\left(\sum_{i=1}^{n} x_i\, y_i\right) - n.\bar{x}.\bar{y}}{\sqrt{\left(\sum_{i=1}^{n} x_i^2\right) - n\bar{x}^2\left(\sum_{i=1}^{n} y_i^2\right) - n\bar{y}^2}}$$

This doesn't look any easier, but it has the advantage of not requiring us to work out any differences. Let's see it by example, as before, with the same data set (so we should get the same answer).

Revision hours	18	2	13	14	6	15	16	9	10	15
Mark	82	20	42	68	41	95	72	48	60	62

We have $\bar{x} = 11.8$ and $\bar{y} = 59$ as before, and also we have $n = 10$, since there are 10 pairs of values here (10 students).

We shall create a table as before, but this will be a little easier. On the top of the fraction we need the sum of all the $x_i\, y_i$, so we work along the columns, multiplying the x_i by the y_i. We start with $18 \times 82 = 1476$, and so on.

On the bottom of the fraction we need the sum of the x_i^2, so we square all of the x_i values (starting from $18^2 = 324$, and so on). We also need the the sum of the y_i^2, so we square all of the y_i values (starting from $82^2 = 6724$, and so on). This is all we need to create for the table, which looks like this:

Revision hours x_i	18	2	13	14	6	15	16	9	10	15	
Mark y_i	82	20	42	68	41	95	72	48	60	62	**Sum**
x_iy_i	1476	40	546	952	246	1425	1152	432	600	930	7799
x_i^2	324	4	169	196	36	225	256	81	100	225	1616
y_i^2	6724	400	1764	4624	1681	9025	5184	2304	3600	3844	39,150

Now we can put all the values into the formula. We have $\sum_{i=1}^{n} x_iy_i = 7799$, $\sum_{i=1}^{n} x_i^2 = 1616$ and $\sum_{i=1}^{n} y_i^2 = 39{,}150$, and remember that $\bar{x} = 11.8$, $\bar{y} = 59$, and $n = 10$.

So, let's calculate the value:

$$r = \frac{\left(\sum_{i=1}^{n} x_iy_i\right) - n.\bar{x}.\bar{y}}{\sqrt{\left(\sum_{i=1}^{n} x_i^2\right) - n\bar{x}^2}\;\sqrt{\left(\sum_{i=1}^{n} y_i^2\right) - n\bar{y}^2}} =$$

$$\frac{7799 - 10 \times 11.8 \times 59}{\sqrt{1616 - 10 \times 11.8^2}\;\sqrt{39{,}150 - 10 \times 59^2}}$$

$$= \frac{837}{\sqrt{223.6}\;\sqrt{4340}} = 0.850 \text{ (to 3 decimal places)}$$

which (thankfully) is the same answer as before.

You can use either version of the formula. The first version is more natural in terms of differences, but this version is probably easier to compute.

> ✔ Do what is right for you. Unless you are specifically told to use a particular formulation of this coefficient, then use whichever version you find easiest, and similarly with other calculations. Check with your lecturer that they don't mind which version you use, though.

● Ranking

There are times when variables don't take actual numerical values, but can be *ranked* in order.

For example, consider a fictional TV show where acts compete, and hope to win over the judges and the audience with their own particular brand of 'talent'. Ten acts have made it through to the final. The two superstar judges will rank the 10 acts in order, from 1 to 10 (1 being first and 10 last), before the audience vote.

Writing the 10 acts in order of performance, the judges voted them from 1 to 10 as follows:

Judge 1	4	3	9	2	8	7	1	10	6	5
Judge 2	6	1	7	3	8	9	2	6	10	4

How closely do the two judges agree with other? We need a measure of 'correlation' that deals with rankings, not actual numerical values.

> **Remember:** These are only rankings, not actual marks. A judge may have thought their first place act was far better than the rest, or they may have thought them all pretty much equal and just put them fairly randomly into some sort of order. There are no values here, only rankings.

● Spearman's rank correlation coefficient

The most commonly used measure of correlation with rankings is *Spearman's rank correlation coefficient*. This was developed by amending Pearson's formula to deal with ranks, and is a good deal simpler. It is often referred to via the symbol ρ, the Greek letter 'rho'.

> Given two sets of ranked data x_i and y_i, Spearman's rank correlation coefficient is given by
>
> $$\rho = 1 - \left| \frac{6\sum_{i=1}^{n} d_i^2}{n(n-1)(n+1)} \right|$$
>
> where $d_i = x_i - y_i$

The only summation here is the top of the fraction, where all we do is take the 'sum of the differences squared'. Let's see this in action on the example above.

Judge 1 (x_i)	4	3	9	2	8	7	1	10	6	5
Judge 2 (y_i)	6	1	7	3	8	9	2	6	10	4
$d_i = x_i - y_i$	-2	2	2	-1	0	-2	-1	4	-4	1
d_i^2	4	4	4	1	0	4	1	16	16	1

In the third row of this table we take the differences between the
judges, and in the fourth row, we square these differences. You've
seen calculations like this before.

If we add up all the values in the last row, we have $\sum\limits_{i=1}^{n} d_i^2 = 51$.

Hence, since in this example $n = 10$ (there are 10 competitors), the
Spearman's rank correlation coefficient is

$$\rho = 1 - \left| \frac{6\sum\limits_{i=1}^{n} d_i^2}{n(n-1)(n+1)} \right| = 1 - \frac{6 \times 51}{10 \times 9 \times 11} = 1 - \frac{306}{990} = 0.691$$

(to 3 decimal places)

This is reasonably high, so it shows that the two judges had a
pretty good (but by no means perfect) agreement between them.
Again, a value of 1 would be a perfect positive correlation, so they
agree completely, and a value of -1 would be a complete negative
correlation, so the judges totally disagree with each other.

● Tied rankings

If a judge decided two rankings were the same (so that two acts
came joint third, for example), the standard technique would be to
give both of these a ranking of 3.5, so the rankings are 1, 2, 3.5,
3.5, 5, 6, 7, 8, 9, 10. Essentially, third and fourth place have been
combined halfway between them, so both have been given 'halfway
between third and fourth place'. You're unlikely to see this much, but
it's worth knowing if you do.

This chapter has been about showing whether one thing is related to another. This is very important in a huge range of subjects: do two things actually have anything in common? Giving a precise mathematical 'answer' to the question enables us to answer this sort of question, which is obviously useful in deciding which techniques to follow, and what to do. (In our main example, revise well, and don't miss classes!)

 Exercises

1 Calculate the Pearson's correlation coefficient of the following data sets. (*Ideally, use both versions of the formula to check that the answers match, and to give you even more practice.*) What can you conclude about the correlation in each question?

(a) The average temperature and the number of visitors per hour to a remote beach, over a number of days:

Temperature	24	12	18	5	7	28	22	20
Visitors	10	6	9	5	4	15	11	12

(b) The number of mistakes made by an employee in a production factory, and the number of months they have worked:

Mistakes	12	3	0	2	9	4	7	20	5	8
Months	2	9	14	12	11	5	5	1	4	6

(c) The height of a footballer, and the number of goals they scored in a season:

Height	1.75	1.84	1.95	1.62	1.79	1.82	1.64	1.72	1.89
Goals	11	3	0	2	5	9	16	8	9

2 The following data ranks the 20 teams in the 2009/10 football Premier League, both in terms of the position in which they finished in the final table, and their ranking in terms of their disciplinary record (from the official Premier League statistics). Use Spearman's rank correlation coefficient to decide whether there is any correlation between the final position ranking and their disciplinary record ranking.

Team	Position rank	Disciplinary rank
Arsenal	3	3
Aston Villa	6	9
Birmingham City	9	15
Bolton Wanderers	14	19
Blackburn Rovers	10	8
Burnley	18	7
Chelsea	1	11
Everton	8	5
Fulham	12	1
Hull City	19	17
Liverpool	7	10
Manchester City	5	2
Manchester United	2	6
Portsmouth	20	18
Stoke City	11	16
Sunderland	13	20
Tottenham Hotspur	4	4
West Ham United	17	13
Wigan Athletic	16	12
Wolverhampton Wanderers	15	14

10 | Linear regression

Creating a linear (straight line) relationship between correlated variables

In the previous chapter we discussed how things could be positively or negatively correlated. If there is a strong correlation, the values should fall roughly in a straight line. What is the best straight line to draw that fits the data most accurately? Can we give a mathematical equation for this line? This is what we shall look at here.

Key topics
- Linear regression
- Gradient and intercept

Key terms
line of best fit regression residuals gradient intercept
method of least squares regression line

● Line of best fit

In the last chapter we talked about correlation, and we discussed how values appear to be correlated if they seem to lie in a straight line. Recall that we had the scattergraph of marks and revision time spent shown in Figure 10.1. There is clearly a positive correlation here, although the points don't fall perfectly in a straight line. Suppose we draw a straight line as best we can, which might look something like Figure 10.2.

This appears to be a reasonable guess at a straight line going through the points. Note that roughly half the values are above the points and half below, and that nothing is too far away. This line may not be perfect, but it's a good guess.

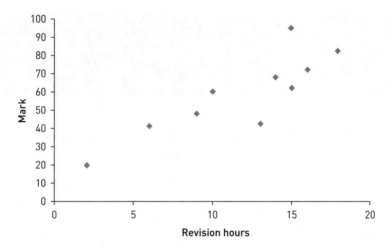

Figure 10.1 **Example of a scattergraph showing hours of revision, forming a positive correlation**

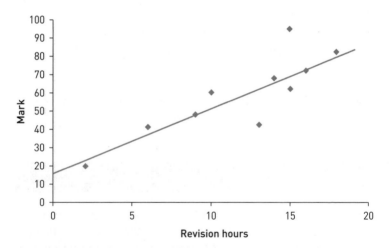

Figure 10.2 **Example of a scattergraph showing hours of revision with a straight line drawn through the positive correlation**

With this data we had a positive correlation (generally speaking, the more the hours of revision, the better the mark), and so the line slopes up. Remember that we also analysed the mark against the number of classes missed, and we had a graph as in Figure 10.3.

A line here might look something like Figure 10.4. This time, we have a negative correlation (generally, the more classes missed, the lower the mark) and the line slopes down.

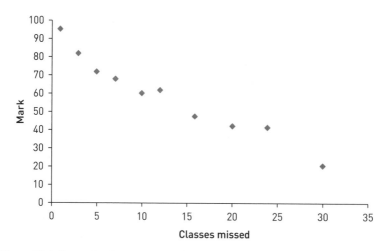

Figure 10.3 **Example of a scattergraph showing the number of classes missed, forming a negative correlation**

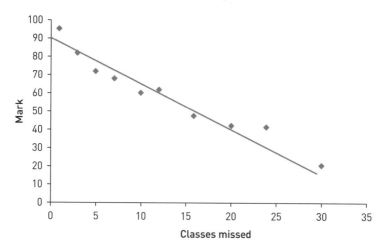

Figure 10.4 **Example of a scattergraph showing the number of classes missed with a straight line drawn through the negative correlation**

This straight line is useful, but so far all I've done is to look at the picture and make a guess as how to draw it. Can we formalise this?

● Gradient and intercept

Before I get into how to define this line for a particular data set, let's discuss a couple of things with regard to straight lines in general.

To define a straight line, we basically need two things. We need to know *where to start*, and we need to know *at what slope* to draw the line. These have formal definitions, which we shall now give.

The *intercept* is the point at which the line crosses the vertical axis. In these examples it is where the line 'begins', so in our first example (the revision hours) it crosses the vertical axis at around 20 (Figure 10.5). Check for yourself that the intercept in the second example (the number of classes missed) is around 90.

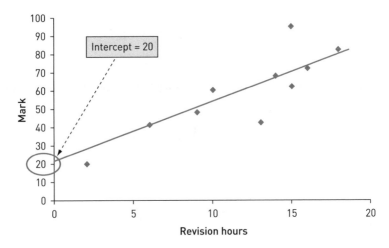

Figure 10.5 Example of a scattergraph showing hours of revision and the intercept point at around 20

The *gradient* is a measure of the 'slope' of the line (what sort of angle it goes up or down at); in fact it is sometimes just referred to as the 'slope'. What you need to know is that a gradient of 0 is perfectly flat (no slope), and as the gradient gets larger, the more the line slopes. If the line slopes down, then the gradient is negative.

For example, a gradient of 2 means that if you go across by 1, you go up by 2, so this is quite a steep slope. A gradient of 0.5 means that if you go across by 1, you go up by 0.5, so this is less steep. Similarly a gradient of −0.5 is a negative (but not particularly steep) slope, whereas −2 is quite a steep negative slope. A gradient of zero doesn't go up or down at all, and so is perfectly flat.

Figure 10.6 illustrates this pictorially. This is not drawn accurately; it's just to give you the idea.

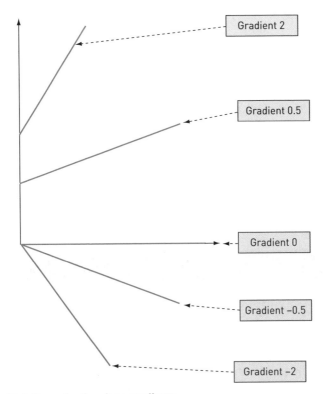

Figure 10.6 **Example of various gradients**

Remember: A positive gradient slopes up, and the more it slopes, the higher the value (so a gradient of 3 is a steeper slope than a gradient of 1). A negative gradient slopes down, and the lower the value, the more it slopes (so a gradient of −3 is steeper than a gradient of −1). A gradient of 0 means that the line is perfectly horizontal.

By convention, the gradient of a straight line is often referred to as m, and the intercept as c. The equation of a straight line can then be given as $y = mx + c$, which means that for any given x-value, the corresponding y-value is obtained by multiplying the x-value by the gradient, and adding the intercept. Remember, what we are trying to do is draw an approximate straight line through all the points.

I'll use m and c, which are quite common, for consistency with the maths book in this series and most maths textbooks, but you may see other notation in the literature, such as a or b, or α and β. It doesn't matter which letters you use, as long as you understand.

If your lecturer, or another book you read, uses a and b, then use that; it means exactly the same thing. If your lecturer asks for the equation in the form $y = ax + b$, then you do exactly the same calculations as you do here where I ask for the equation in the form $y = mx + c$.

✔ Different people do use different notation, so you need to be aware of the notation being used. Some people like to use m and c for gradient and intercept, while others prefer to use a and b. It does not matter at all; all notation is just letters to stand for something. Just make sure you are aware of the particular notation being used.

● The method of least squares

There are various ways to work out the 'best' line to use, but the simplest one is probably the *method of least squares*. The reasoning behind this is similar to the reasoning behind the variance. Essentially, we want to draw a line such that the overall distance between the points and the line is as small as possible. However, we want to make sure that the values of distance that we work with are all positive, so that they don't cancel out, and we also want to avoid having any values far away. So, exactly as we did for variances, taking the square of the distances deals with both of these problems: squares are always positive, and values further away will make a greater contribution. The distance between a point and the line is known as a *residual*.

With this in mind, we can create a formula for the line. The proof of this formula is beyond what I want to cover here, but you can see that it involves taking differences between values and the mean.

● The linear regression formula

As before, I shall give the formula, and then do some examples to try to make the use of the formula clear.

Given a set of n pieces of data x_i and y_i, with mean $\bar{x}$ and $\bar{y}$, the equation of the regression line is given by $y = mx + c$, where

$$m = \frac{\displaystyle\sum_{i=1}^{n}(x_i - \bar{x})(y_i - \bar{y})}{\displaystyle\sum_{i=1}^{n}(x_i - \bar{x})^2}$$

and $c = \bar{y} - m\bar{x}$

This is quite similar to formulae we have used before, and we can attack it in a similar way, making tables to help us calculate. It's much easier to see this when we do an example.

● Linear regression: Example 1

Let's do the same example as before, where we compared the marks obtained with the revision hours. This data is repeated below:

Revision hours (x_i)	18	2	13	14	6	15	16	9	10	15
Mark (y_i)	82	20	42	68	41	95	72	48	60	62

Recall we worked out that the means were $\bar{x} = 11.8$ and $\bar{y} = 59$.

We can calculate a table almost exactly as in the previous chapter; the only difference is that we don't need to work out the $(y_i - \bar{y})^2$ terms.

Revision hours (x_i)	18	2	13	14	6	15	16	9	10	15	
Mark (y_i)	82	20	42	68	41	95	72	48	60	62	**Sum**
$x_i - \bar{x}$	6.2	−9.8	1.2	2.2	−5.8	3.2	4.2	−2.8	−1.8	3.2	0
$y_i - \bar{y}$	23	−39	−17	9	−18	36	13	−11	1	3	0
$(x_i - \bar{x})(y_i - \bar{y})$	142.6	382.2	−20.4	19.8	104.4	115.2	54.6	30.8	−1.8	9.6	837
$(x_i - \bar{x})^2$	38.44	96.04	1.44	4.84	33.64	10.24	17.64	7.84	3.24	10.24	223.6

Now we can work out $m = \dfrac{\displaystyle\sum_{i=1}^{n}(x_i - \bar{x})(y_i - \bar{y})}{\displaystyle\sum_{i=1}^{n}(x_i - \bar{x})^2} = \dfrac{837}{223.6}$ directly from

this table. From this, we have $c = \bar{y} - m\bar{x} = 59 - \dfrac{837}{223.6} \times 11.8 = 14.829$ (to 3 decimal places).

Note that we did not round off $\frac{837}{223.6}$ for the calculation, so as not to introduce errors, but now we are presenting the final answer we can do so; it works out to be 3.743.

So, we have $m = 3.743$ and $c = 14.829$.

Therefore the equation of the regression line is $y = 3.743x + 14.829$. This means that it has an intercept of 14.829 (where it crosses the vertical axis) and a slope of 3.743.

If you draw a graph of this line, on the same graph that we used for the scattergraph, you will see that we have what looks to be a good-fitting line (Figure 10.7). It's better than my 'do-it-by-guessing' line before. Look back at that line; it's not bad, but the intercept is a fair way out, I guessed it was about 20, but the maths says it is 14.829.

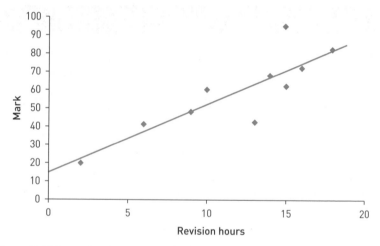

Figure 10.7 **Example of a scattergraph showing the intercept point at 14.829 and a slope of 3.743**

● Linear regression: Example 2

Let's do another example, this time with negative correlation. We shall use the marks obtained and the classes missed from before.

Classes missed	3	30	20	7	24	1	5	16	10	12
Mark	82	20	42	68	41	95	72	48	60	62

Recall that the graph looked like Figure 10.8, and we calculated that there was a very strong negative correlation. You can calculate that

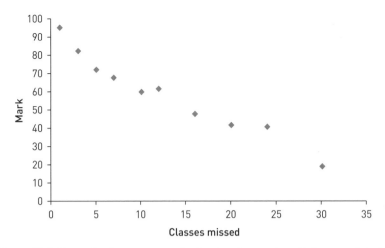

Figure 10.8 Example of a scattergraph showing the number of classes missed, forming a negative correlation

the mean of the x_i is $\bar{x} = 12.8$, and the mean of the y_i is $\bar{y} = 59$. As before, we can create the following table:

Missed (x_i)	3	30	20	7	24	1	5	16	10	12	
Mark (y_i)	82	20	42	68	41	95	72	48	60	62	**Sum**
$x_i - \bar{x}$	−9.8	17.2	7.2	−5.8	11.2	−11.8	−7.8	−3.2	−2.8	−0.8	0
$y_i - \bar{y}$	23	−39	−17	9	−18	36	13	−11	1	3	0
$(x_i - \bar{x})(y_i - \bar{y})$	−225.4	−670.8	−122.4	−52.2	−201.6	−424.8	−101.4	−35.2	−2.8	−2.4	−1839
$(x_i - \bar{x})^2$	96.04	295.84	51.84	33.64	125.44	139.24	60.84	10.24	7.84	0.64	821.6

Using the regression formula, this gives

$$m = \frac{\sum_{i=1}^{n}(x_i - \bar{x})(y_i - \bar{y})}{\sum_{i=1}^{n}(x_i - \bar{x})^2} = \frac{-1839}{821.6}$$

directly from this table. From this, we have

$$c = \bar{y} - m\bar{x} = 59 - \frac{-1839}{821.6} \times 12.8 = 87.650$$

(to 3 decimal places)

Rounding off $\frac{-1839}{821.6}$ to be −2.238 for the final answer (remember, we didn't round it before, so as not to introduce errors in the calculation), we have $m = -2.238$ and $c = 87.650$.

Therefore the equation of the regression line is $y = -2.238x + 87.650$. This means that it has an intercept of 87.650 (where it crosses the vertical axis) and a slope of -2.238.

If you draw a graph of this line, on the same graph that we used for the scattergraph, you will see that we have what looks to be a good-fitting line (this time, not too far away from my 'guessing by looking at it' line from before) (Figure 10.9).

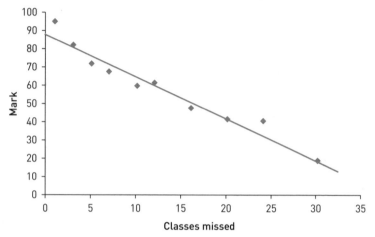

Figure 10.9 **Example of a scattergraph showing the intercept point at 87.650 and a slope of -2.238**

● Use of the regression line

The linear regression line is used because it is straight, and straight lines are quite easy to deal with – much easier than complicated curves. If we have a strong correlation, then we can approximate our data by just using the regression line. So, for example, in our list of classes missed and marks obtained, the line was $y = -2.238x + 87.650$. How can we use this?

For example, we can use it to tell us that a student who missed no classes can expect a mark of $-2.238 \times 0 + 87.650 = 87.650$, a student who missed 8 classes (this is not a piece of data in the original set) can expect a mark of $-2.238 \times 8 + 87.650 = 69.746$, and so on. Care has to be taken not to go too far beyond the original set. If you try to work out the expected mark of a student missing 50 classes, for example, you'll get a negative answer, so the calculation isn't valid (you can't get a negative mark!). There probably weren't even 50 classes to miss, so the calculation doesn't make sense.

This means that from the data we have, we have a simple formula to use. Probably the moral out of all of this is to attend all your classes!

● Further regression

You can of course develop deeper models of regression, which can be more accurate than a simple straight line, such as a curved line, for example. But this brings in its own complexity and difficulty, as the equations for curved lines are much harder than those for straight lines. This is a subject that is still developing today and forms the basis for much statistical research; linear regression is only a start.

> Never forget that subjects are growing and learning; nobody knows everything. There is current active research in making things like better regression techniques, and this is going on in universities right now. Don't think that your lecturers know everything, and that their job is just to tell you everything; subjects are evolving and developing all the time.

Summary

In complex cases there is a lot of data, and we wish to approximate it to make a simple calculation. Where the correlation is strong, then the straight line is quite close to the original points, and straight lines are quite easy to deal with. So we choose to deal with the straight line rather than the complex data set. Our answers may be a little out, but it's all probabilistic anyway, and we're probably close enough to make our answer worthwhile.

> **Remember:** Complex statistics is too hard and time-consuming to analyse, and so approximations need to be made. What we are doing here is taking something hard and putting an 'easy' approximation (in this case a straight line) on it. This saves huge amounts of time and effort, and if the correlation is strong, the line is pretty close to the actual data, so our answers are pretty good. For most of the really complex problems you can encounter in statistics, an 'easy, fast, good' solution is perfectly acceptable, and much faster (and so certainly cheaper for a company trying to do it) than a perfectly precise solution; this is what this regression line does for us.

Exercises

1 For the examples in Chapter 9, given again below, calculate the equation of the regression line.

(For full understanding and practice I suggest you first draw a line of best fit on a scattergraph by eye – just the best fit you can create by looking at the graph. Then calculate the regression line and draw that, and see how they compare.)

(a) The average temperature and the number of visitors per hour to a remote beach, over a number of days:

Temperature	24	12	18	5	7	28	22	20
Visitors	10	6	9	5	4	15	11	12

(b) The number of mistakes made by an employee in a production factory, and the number of months they have worked:

Mistakes	12	3	0	2	9	4	7	20	5	8
Months	2	9	14	12	11	5	5	1	4	6

(c) The height of a footballer, and the number of goals they scored in a season:

Height	1.75	1.84	1.95	1.62	1.79	1.82	1.64	1.72	1.89
Goals	11	3	0	2	5	9	16	8	9

2 Similarly to the correlation formula, there is an alternative formula for the regression line, which is given by

$$\text{Gradient } m = \frac{\left(\sum_{i=1}^{n} x_i y_i\right) - n.\bar{x}.\bar{y}}{\left(\sum_{i=1}^{n} x_i^2\right) - n\bar{x}^2} \text{ and intercept } c = \bar{y} - m\bar{x}$$

Use this formula to work out the regression line for the two examples in the chapter (and the three questions above if you wish), and verify that you get the same answer.

PROBABILITY

An introduction to probability

Working out probabilities: the chances of something happening

Many things in life come down to luck or chance. You can't know for certain what will happen with most things, but you can know that one thing is more likely than another. How can we mathematically identify the likelihood of something happening? This is the essence of probability.

Key topics
- Probability
- Calculating discrete probabilities

Key terms
probability events complement

A lot of what happens in life is down to chance, and randomness plays a huge role in statistics. Probability is all about mathematically calculating the chance, or likelihood, of something happening. At least if we know how likely something is to happen, we can take some measure to predict whether it is likely, and make some plans to deal with it. In this chapter we shall introduce probability in an informal way.

● Probability

Before we define probability mathematically, let's think about what it 'should' mean. Take, for example, a normal die ('die' is the singular of 'dice') with six sides, numbered 1, 2, 3, 4, 5 and 6.

Suppose you throw this die randomly. What are the 'chances' that you will throw a 5? There are six possibilities – you might get a 1, a

2, a 3, a 4, a 5 or a 6 – and each one is equally likely. So the chances of getting a 5 might be seen as '1 in 6' – basically, 1 out of the 6 equally likely possibilities. You can write this as a fraction: the probability of throwing a 5 is $\frac{1}{6}$.

What is the chance that you will score less than 6? Out of the six possible outcomes, all of them are less than 6, apart from 6 itself. So if you get 1, 2, 3, 4 or 5, then you have scored less than 6. So five out of the six possibilities mean you will score less than 6, and so the probability of scoring less than 6 should be $\frac{5}{6}$.

● Discrete probabilities

Following on from the above example, we can generalise the concept of the probability of an *event* (a specific happening) occurring. For now, we shall look only at what are called *discrete* probabilities: that is, ones where there are a fixed number of distinct possible outcomes.

The definition of the probability of an event occurring is given by:

$$P(event) = \frac{\text{the number of outcomes in which that event can occur}}{\text{the total number of possible outcomes}}$$

It might not be immediately obvious what this means, so let's illustrate with an example.

Suppose we are picking a random integer (whole number) between 1 and 10. Hence there are 10 possible outcomes: 1, 2, 3, 4, 5, 6, 7, 8, 9 or 10.

We are going to calculate the probability (chance) of picking the following. Note that we can give probabilities either as fractions or decimals, but you should generally use decimals only when they are exact answers (not rounded), or you risk introducing rounding errors.

(a) 7

(b) A number less than 7

(c) An even number (divisible by 2)

(d) An odd number (not divisible by 2)

(e) Anything apart from 7

(f) At least 1

(g) 20

Solutions:

(a) Only one of the ten possible outcomes gives 7, and so
$$P(\text{picking } 7) = \frac{1}{10} = 0.1.$$

(b) There are six out of the ten possible outcomes that give an answer that is less than 7 (namely 1, 2, 3, 4, 5 and 6), and so
$$P(\text{picking a number less than } 7) = \frac{6}{10} = \frac{3}{5} = 0.6.$$

(c) There are five out of the ten possible outcomes that give an answer that is even (namely 2, 4, 6, 8, and 10), and so
$$P(\text{picking an even number}) = \frac{5}{10} = \frac{1}{2} = 0.5.$$

(d) There are five out of the ten possible outcomes that give an answer that is odd (namely 1, 3, 5, 7, and 9), and so
$$P(\text{picking an odd number}) = \frac{5}{10} = \frac{1}{2} = 0.5.$$

(e) Nine out of the ten possible outcomes give an answer that is not 7 (namely 1, 2, 3, 4, 5, 6, 8, 9 and 10), and so
$$P(\text{picking a number other than } 7) = \frac{9}{10} = 0.9.$$

(f) All ten of the possible outcomes are at least 1, and so
$$P(\text{picking a number that is at least } 1) = \frac{10}{10} = 1.$$

(g) Finally, none of the ten possible outcomes can give 20 – it's impossible to obtain as an outcome – and so $P(\text{picking } 100) = \frac{0}{10} = 0.$

Note in the last two examples that if our desired outcome is certain, we have probability 1, and if the desired outcome is impossible, we have probability 0.

> Probabilities can also be given as percentages. For example, a probability of 0.4 is the same as 40%, and so we might talk about a 40% probability. We shall just use fractions and decimals here, but when we get deeper into statistics later you will see more use of percentages, so make sure you are comfortable with all of fractional, decimal, and percentage notation.

Complements

Given an event, the *complement* of that event is essentially the 'complete opposite' of it.

For example, what is the complement (opposite) of picking 7 in the above example? It is those outcomes that mean we *don't* pick 7. The outcomes that make these true are as follows:

Picking 7: outcome of 7
Not picking 7: outcome of 1, 2, 3, 4, 5, 6, 8, 9 or 10

Note that every outcome appears once in these lists: either in the first list or in the second one.

If you add up the probability of picking 7, and the probability of its complement (not picking 7), you get the answer 1:

$$P(\text{picking 7}) + P(\text{not picking 7}) = \frac{1}{10} + \frac{9}{10} = 1$$

Similarly, what is the complement of picking an even number? It's the complete opposite: you don't pick an even number, or in other words, you pick an odd number.

Even number: outcomes of 2, 4, 6, 8, 10
Odd number: outcomes of 1, 3, 5, 7, 9

Note, again, that every possible outcome appears once in these lists, either in the first list or in the second list. And again note that the sum of the probabilities is 1:

$$P(\text{even number}) + P(\text{odd number}) = \frac{5}{10} + \frac{5}{10} = 1$$

This rule is true in general:

> If you add together the probability of an event, and the probability of its complement, you get 1.

Another example

You need to be very careful with probabilities and ensure that you consider all possible outcomes. For example, consider what happens when you toss two coins. What is the probability that you will get:

(a) two heads,

(b) one head and one tail?

You have to be very careful here. What are the possible outcomes?

Intuitively a lot of people would say that there are three possible outcomes (two heads, two tails, or a head and a tail), and so both probabilities are $\frac{1}{3}$. **This is not correct!**

If you consider the outcome of the first coin and the second coin separately, after some thought you will conclude that there are four possible outcomes:

First coin	Second coin
HEAD	HEAD
HEAD	TAIL
TAIL	HEAD
TAIL	TAIL

Hence the answers to the questions are:

(a) The only one of the four possible outcomes that gives us two heads is the first row of the table, so $P(\text{two heads}) = \frac{1}{4}$.

(b) In both the second and third rows of the table, we have one head and one tail. Hence $P(\text{a head and a tail}) = \frac{2}{4} = \frac{1}{2}$.

The important thing here is that 'a head and a tail' is interpreted as being the same as 'a tail and a head', and so it's twice as likely to happen as getting two heads.

> ✔
>
> Experiment for yourself. If you aren't convinced by the above, for example, then take two coins and throw them yourself a good number of times (say 50) and you should find that around half the time you get a head and a tail. Roll a die 50 times and you'll probably get roughly half even and half odd. Maybe you can come up with your own games? Make maths practical and fun and it will be much easier for you to appreciate it.

↻ Summary

This chapter is a very brief introduction to probability – we shall study it in more detail in the next few chapters – but hopefully you have grasped the idea.

Remember that probability gives us a measure of what the chance or likelihood of an event happening is. If we roll a die, we can expect that, one in every six times, we'll roll a 6.

This *doesn't* mean that if you roll a die six times, you are guaranteed to roll a 6. It means that in the long run, on average, you'll get a six once every 6 rolls. Sometimes you might go on a great run and get three 6s in a row – but other times it might seem like an eternity before you throw a 6. Remember, it's all about what you 'expect' to happen.

✎ Exercises

1 A normal die (taking values 1 to 6) is rolled. What is the probability that you will roll the following? Leave your answers as fractions in their lowest terms where appropriate.

Example: A number greater than 4.

Solution: Of the six possibilities (1, 2, 3, 4, 5, 6) then only two of these (5, 6) are greater than 4, so the probability is $\frac{2}{6} = \frac{1}{3}$.

(a) 4
(b) 1 or 6
(c) An even number
(d) An integer less than 10
(e) Not 2
(f) A negative number
(g) Either 2 or an odd number
(h) A number divisible by either 2 or 3
(i) Either a number less than 3 or a number greater than 3

2 A random letter of the English alphabet (26 letters) is chosen. What is the probability that the following will be chosen? Leave your answers as fractions in their lowest terms where appropriate.

Example: Either A or B

Solution: Of the 26 possibilities, only two are A or B, so the probability is $\frac{2}{26}$ $= \frac{1}{13}$.

(a) Q

(b) A, B, C or D

(c) A vowel (A, E, I, O, U)

(d) A consonant (not a vowel)

(e) A letter earlier in the alphabet than N

3 If you toss three coins, there are eight possibilities. Can you write down the eight possibilities (try to be systematic)?

Using your answer, calculate the following:

(a) If you toss three coins, what is the probability of getting three tails?

(b) If you toss three coins, what is the probability of getting two heads and one tail?

4 This is known as the Monty Hall problem, after a US television programme from the 1960s/1970s. In a game show, a contestant has three doors in front of them. Behind one is the star prize of a car, and behind the other two are unwanted prizes of goats. Only the game show host knows which door the car is behind. The contestant chooses a door at random, and the host opens one of the other doors to reveal a goat. The contestant is now offered the chance to switch from their original choice. Should they switch or not, or does it not matter?

(Hint: write down all the possibilities for how the objects are placed behind the doors, and then consider in each case what happens if you keep your original choice or switch, and work out the probability of winning.)

Probabilities when more than one event is involved

Probabilities get more complicated when there is more than one event involved. For example, what happens when you toss a coin and roll a die? In this chapter we shall look at ways to solve problems in probability when more than one event is involved in the calculation.

Key topics

● Probability of both one event and another
● Probability of either one event or another

Key terms
and or independent mutually exclusive

Working out probabilities of more complex events can seem quite tricky, but there are fairly straightforward ways to go about it. The main thing you need to concern yourself with is what the possibilities are, and whether the outcome of one event has any impact on the outcome of another.

● 'And'

Suppose you both tossed a coin, and rolled a die. What is the probability that:

(a) the coin is tails and the die rolls 1,

(b) the coin is heads and the die rolls an odd number?

This is somewhat trickier. There are many combinations that give possible outcomes. The coin can either be a head or a tail, and the die can roll 1, 2, 3, 4, 5 or 6.

In total there are 12 possible outcomes: writing them as pairs (coin toss, die roll) they are:

(head, 1), (head, 2), (head, 3), (head, 4), (head, 5), (head, 6), (tail, 1), (tail, 2), (tail, 3), (tail, 4), (tail, 5), (tail, 6).

So for (a) the only one of these that satisfies our requirement is the pair (tail, 1), and so our probability is $\frac{1}{12}$ (one of the 12 possible outcomes).

For (b) there are three possible outcomes that are meet our requirement: (head, 1), (head, 3) and (head, 5). So our probability is $\frac{3}{12}$ (three of the 12 possible outcomes), which is $\frac{1}{4}$.

Now, note the following:

In (a), the probability of a tail is $\frac{1}{2}$, and the probability of rolling a one is $\frac{1}{6}$. If you multiply these together you get $\frac{1}{2} \times \frac{1}{6} = \frac{1}{12}$, which is the probability that you just worked out of both events happening together (both a tail and a one).

In (b), the probability of a head is $\frac{1}{2}$, and the probability of rolling an odd number is $\frac{3}{6} = \frac{1}{2}$. If you multiply these together you get $\frac{1}{2} \times \frac{1}{2} = \frac{1}{4}$, which is the probability that you just worked out of both events happening together (both a head and an odd number).

This works in general as long as the two events are *independent*, that is, they have no connection with each other (the tossing of the coin does not affect the roll of the die, and vice versa, so the events are independent).

We often use the symbol ∩ for the idea of 'and'. Using the usual probability notation, then, we can say:

$$P(A \cap B) = P(A) \times P(B)$$

If you don't want to use this symbol (known as the *intersection symbol* after ideas from set theory), then you can just write this as $P(A \text{ and } B) = P(A) \times P(B)$, but be aware of the symbol usage. To write this in words:

> The probability of one event *and* another event happening is obtained by *multiplying* the probabilities together, *as long as the events are independent.*

● 'Or'

Sometimes questions are asked where the desired outcome can be expressed in the form 'either one thing, or something else'. For example, what is the probability, when you roll a die, that the outcome will be either 6, or an odd number?

There are six possible outcomes, of which four are the desired outcome: 6 (which satisfies the first condition) and 1, 3, 5 (which satisfy the second condition). So P(either 6, or odd) $= \frac{4}{6} = \frac{2}{3}$.

Now, note the following:

$$P(\text{rolling } 6) = \frac{1}{6}$$

$$P(\text{an odd number}) = \frac{3}{6} = \frac{1}{2}$$

If you add up these probabilities, you get $\frac{1}{6} + \frac{1}{2} = \frac{4}{6} = \frac{2}{3}$, which is the same answer as we had above.

Does this always work? **No!**

It works only when the two events are *mutually exclusive* (or *disjoint*) – which means they have nothing in common, so they cannot both happen. This is fine in this example, as 6 is not odd, and so you cannot meet both of the criteria (it cannot be both 6 and odd).

 Remember: Two events are mutually exclusive if they cannot *both* happen.

Similarly to the way we used ∩ for 'and', it is common practice to use the symbol ∪ for 'or' (this is called the *union* symbol, again from set theory).

Hence we can say:

If A and B are mutually exclusive events, then $P(A \cup B) = P(A) + P(B)$

Again, if you don't want to use the union symbol, then you can just write this as $P(A \text{ or } B) = P(A) + P(B)$. To write this in words:

> The probability of one event *or* another event happening is obtained by *adding* the probabilities together, *as long as the events are mutually exclusive.*

As an example of where this doesn't work, suppose I roll the die again; what is the probability that I will get either a number more than 4, or an even number?

The outcomes that make this work are either 5, 6 (numbers more than 4), or 2, 4, 6 (even numbers). So, in total, a roll of 2, 4, 5 or 6 meets our requirement, four possibilities out of the six. So the probability is $P(\text{more than 4, or even}) = \dfrac{4}{6} = \dfrac{2}{3}$.

But $P(\text{more than 4}) = \dfrac{2}{6} = \dfrac{1}{3}$ and $P(\text{even}) = \dfrac{3}{6} = \dfrac{1}{2}$, and if you add these up you get $\dfrac{1}{3} + \dfrac{1}{2} = \dfrac{5}{6}$, which is NOT the right answer.

The problem comes because the number 6 is in both of the two possibilities, and so the two events are not mutually exclusive: it is possible for both of them to happen (if I roll a 6).

Essentially, when counting the possibilities, we have counted the 6 twice (once for each event). In general, we can add up the two probabilities, but then we need to take away the probability of both events happening, to compensate for the fact that we counted these things twice.

Therefore, if A and B are not mutually exclusive (it is possible for them both to happen), then:

$$P(A \cup B) = P(A) + P(B) - P(A \cap B)$$

Note that this formula is consistent with the first formula (when the events are mutually exclusive), since if they are mutually exclusive, both A and B cannot happen, and so $P(A \cap B) = 0$, and we just get $P(A \cup B) = P(A) + P(B)$ as before.

To illustrate this formula, take exactly the same problem as before: if we roll a die, what is the probability that we will get either a number more than 4, or an even number?

If A is the event 'rolling more than 4', and B is the event 'rolling an even number', then $P(A) = \dfrac{1}{3}$ and $P(B) = \dfrac{1}{2}$ as before, and $P(A \cap B) = \dfrac{1}{6}$ (since there is only one of the six possibilities that satisfies A and B, namely if you roll a 6).

Hence $P(A \cup B) = P(A) + P(B) - P(A \cap B) = \frac{1}{3} + \frac{1}{2} - \frac{1}{6} = \frac{2}{3}$

(check this for yourself), which is what we calculated to be the right answer by thinking of all the possibilities.

> ✔ Learn these formulae by concept. If you just remember that 'and' is multiplying and 'or' is adding, you've essentially remembered the formulae, as long as you remember the 'mutually exclusive' bit. This is far easier to learn that the actual symbols 'P bracket A union B bracket equals ...', and if you learn it conceptually, you also understand it, and understanding is far more important than mere knowledge.

● Independence and mutual exclusivity

In this chapter you have seen both the terms *independent* and *mutually exclusive*. I want to remind you that these two things have very different meanings.

Independence is to do with multiplication of probabilities: two events are independent if $P(A \cap B) = P(A) \times P(B)$.

Mutual exclusivity is to do with addition of probabilities: two events are mutually exclusive if $P(A \cup B) = P(A) + P(B)$.

For example, if you pick a card from a standard deck of playing cards, consider the events $A = $ 'red card' and $B = $ 'an ace'. These two events are independent, since $P(A) = \frac{1}{2}$ (half of the cards are red), $P(B) = \frac{1}{13}$ (there are four aces in the pack of 52 cards, and $\frac{4}{52} = \frac{1}{13}$), and $P(A \cap B) = \frac{2}{52}$ (since there are two red aces) $= \frac{1}{26}$. Hence we do have $P(A \cap B) = P(A) \times P(B)$ (check this for yourself), and so the events are independent.

But they are not mutually exclusive: it is possible for a card to be both red and an ace. As we pointed out, there are two such cards, the ace of hearts and the ace of diamonds. It is possible for the two events to be mutually exclusive only if $P(A \cap B) = 0$, and we have just worked out this is not the case.

Please don't get the two concepts confused; treat them as entirely separate things, so that there is no danger of confusing the terms!

There are quite a few terms here, and unfortunately statistics is rather full of words and terms that won't be immediately familiar to you (like 'mutually exclusive'). But do learn them, and what they mean, as you *will* encounter them, and you'll want to be able to remember what they mean without having to look them up every time. Use the glossary in this book to help you at first (and keep it handy), until the terms become more familiar and eventually form part of your vocabulary.

Summary

In this chapter we have taken probability a step forward and investigated the probability of multiple events happening. Of course, you can generalise this further to three events and more, and higher-level statistics will do this, but this is enough for an introduction – it introduces the concept, and that is the most important thing for an introductory textbook like this. Hopefully you now understand the concepts of 'and' and 'or' in probability, so that you can answer problems that arise.

Exercises

1 A player rolls a die, and also independently randomly selects a card from a pack containing the integers between 1 and 10 inclusive. What is the probability of the following?

Example: They roll a 6 and select an even numbered card.

Solution: The probability of rolling a 6 is $\frac{1}{6}$ and the probability of selecting an even number is $\frac{5}{10} = \frac{1}{2}$. Since this is an 'and' problem, we need to multiply the probabilities, and we get the answer $\frac{1}{6} \times \frac{1}{2} = \frac{1}{12}$.

(a) They roll a 1 and also select card 1.

(b) They roll an even number and select an even numbered card.

(c) They roll an odd number and select a card with value 9 or 10.

(d) They roll more than 1 and select a card with value more than 1.

(e) Both numbers are less than 3.

(f) The two numbers are the same. (*Hint: this is easier than it seems – does it actually matter what they roll on the die?*)

2 A bag of children's toys contains 3 red circles, 2 green triangles, 2 green squares, 2 red squares and 1 red ball, from which a child selects one item at random.

Example: They select either a square or something red.

Solution: The events are not mutually exclusive as there are 2 red squares. Using the formula $P(\text{square or red}) = P(\text{square}) + P(\text{red}) - P(\text{square and red})$, you get $\frac{4}{10} + \frac{6}{10} - \frac{2}{10} = \frac{8}{10} = \frac{4}{5}$

(a) They select a square or something green.

(b) They select something red or a triangle.

(c) They select a ball or a square.

(d) They select something red, or a green square.

Probability trees

Multiple probabilities expressed in a probability 'tree'

It can be easier to express multiple probabilities in a diagrammatical structure. These structures are referred to as trees, and can be very useful in calculating probabilities in a way that is clear and visually appealing

Key topics
- Probability trees

Key terms
probability trees paths

Often a problem can change as we work through it: for example, when playing a game, what we do after the first step might depend on what happened to us in that first step. Although you can deal with this completely mathematically, with conditional and multiple probabilities as in the last chapter, this can get messy and awkward. Presenting things pictorially can really help make it clear what you are doing, which lessens the chance of mistakes, and makes it easier for someone else to understand your work.

We shall present this chapter as a fully worked example.

● Rules of the game

The rules of a game involving a standard die (numbers 1 to 6), a coin (heads or tails) and a standard pack of cards (52 cards, 4 each of ace, two, three, four, five, six, seven, eight, nine, 10, jack, queen, king) are as follows:

> First you roll the die. If you roll a 6, then you pick a card, and as long as it is not an ace you win, otherwise you lose. If you don't roll a 6, you toss a coin, and if it's heads, you win, and if it's tails, you lose.

Is this game fair? What are your chances of winning? Before we get into the mathematical calculation, think about it for a while. Do you think you are more likely to win or lose?

● Creating probability trees

We can illustrate this game by a 'tree' structure. Start from the beginning, when you roll the die. There are two possibilities: either you roll a 6, or you don't. The probability of rolling a 6 is $\frac{1}{6}$ and the probability of not rolling a 6 is $\frac{5}{6}$.

So this could be viewed as in Figure 13.1, where we have included the probabilities. You can view this picture as representing our two options: either you roll a 6 or you don't. Depending on what happens, you have to take one of the two paths. If you get a 6, you follow the top path; if you don't, you follow the bottom path.

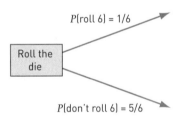

Figure 13.1 **Example of a probability tree**

The important point is what happens next; what you do depends on what path you took.

First, suppose you roll a 6. Then you pick the card; if it's anything other than an ace you win. So your chances of winning are $\frac{48}{52}$ (any of the cards apart from the four aces), which is $\frac{12}{13}$. Your chances of losing are $\frac{4}{52} = \frac{1}{13}$ (you pick one of the four aces).

Remembering that, for now, we are considering just what happens if you follow the first path (so you roll a 6 with the die), our diagram looks like Figure 13.2.

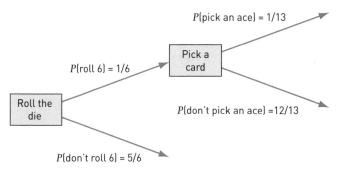

Figure 13.2 **Example of a probability tree showing the top path (a 6 is rolled)**

Now consider what happens if you follow the second path (you don't roll a 6). Then you have to toss the coin, which can be either heads (probability $\frac{1}{2}$) or tails (probability $\frac{1}{2}$). So now your diagram is as in Figure 13.3.

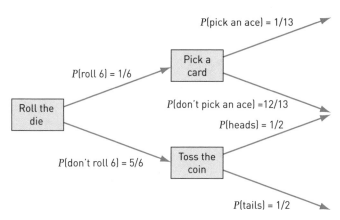

Figure 13.3 **Example of a probability tree showing both the top and bottom paths**

Now we have four paths. Let's work out the probability of following each of the four paths. Remember our rule for multiple probabilities: we need everything on the path to happen, so it is an 'and' question. Therefore we need to multiply the probabilities together. Let's see it by example on each of the four paths:

The first path: you roll a 6 and then pick an ace.
The probability of this happening is $\frac{1}{6} \times \frac{1}{13} = \frac{1}{78}$ (= 0.013 (3dp)) (dp is short for 'decimal places'). If it does happen, you lose.

The second path: you roll a 6 and then pick anything other than an ace.

The probability of this happening is $\frac{1}{6} \times \frac{12}{13} = \frac{12}{78} = \frac{2}{13}$ (= 0.154 (3dp)). If it does happen, you win.

The third path: you don't roll a 6 and then toss heads.

The probability of this happening is $\frac{5}{6} \times \frac{1}{2} = \frac{5}{12}$ (= 0.417 (3dp)). If it does happen, you win.

The fourth path: you don't roll a 6 and then toss tails.

The probability of this happening is $\frac{5}{6} \times \frac{1}{2} = \frac{5}{12}$ (= 0.417 (3dp)). If it does happen, you lose.

So now we can extend our tree diagram to include the overall probabilities and outcome of each path we can take (Figure 13.4). So there are two paths by which you win, and two paths by which you lose. What is the probability of you winning?

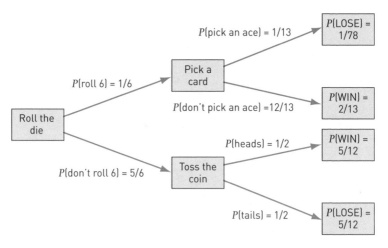

Figure 13.4 Example of a probability tree showing all probabilities and outcomes

To win, we have to go along one of the two paths that ends at a win – the second path and the third path. Because these paths are different, this is an 'or' question (either the second or the third path) where the paths are mutually exclusive, so we can add the two probabilities to work out the overall probability that we shall end in a winning position.

Expressing this more mathematically:

P(winning) = P(following the second path) + P(following the third path)

(since these are the two paths that lead us to a win)

$$= \frac{2}{13} + \frac{5}{12} = \frac{89}{156} = 0.571 \text{ (3dp)}$$

Also,

$P(\text{losing}) = P(\text{following the first path}) + P(\text{following the fourth path})$

(since these are the two paths that lead us to a loss)

$$= \frac{1}{78} + \frac{5}{12} = \frac{67}{156} = 0.429 \text{ (3dp)}$$

Note that these two probabilities add up to 1, as they should; you either win or lose.

Hence your chances of winning are about 0.571, and your chances of losing are about 0.429. This means that, in the long run, you can expect to win more than half of the time, and so this game is definitely worth playing.

Note, by the way, that all paths don't have to be the same length, as long as they finish. For example, suppose that in the above game, if you roll a 6, you just win automatically without needing to pick a card. Then our diagram would look like Figure 13.5.

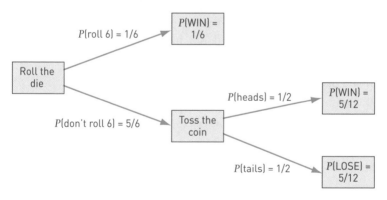

Figure 13.5 Example of a probability tree showing the outcome of rolling a 6 and winning automatically

Now your chances of winning are $\frac{1}{6} + \frac{5}{12} = \frac{7}{12} = 0.583$ (3dp), which is slightly higher, as you would expect, since you don't have to worry about possibly getting an ace with the choice of card if you are lucky enough to roll a 6.

The reason why these diagrams are referred to as *trees* is that you can visualise them as branches growing from a trunk, just like a tree.

This entire chapter is a worked example. Often, learning by
example is the best way to get to grips with what something is all
about. When you learn a topic, try to work through examples (create
your own if you can) to reinforce your confidence in what you are doing,
and the reason for doing it.

 Summary

Probability trees are an easy way to see what is actually going on
when lots of events are happening. Looking at things visually is
often really helpful, as it breaks the problem down into the relevant
parts, and you can see exactly what is happening at each stage. One
of the best tools you can have is a pen and paper, to visualise and
describe what is going on.

Exercises

1 In a game show, you first have to pick a ball from a bag
containing three balls: one red, one blue, and one green. You are
then offered a choice from four boxes:

- If you picked the red ball, only one box has a winning ticket,
- If you picked the green ball, two boxes have winning tickets,
- If you picked the blue ball, three boxes have winning tickets,

Draw a tree to represent this game. What is the overall probability
that you will win?

2 I challenge you to a game. You roll a die, and if you get a 6 you
immediately win. Otherwise I ask you to roll another die: if you get a
5 or a 6 this time you win, otherwise you lose.

Draw a tree to represent this game. What are the chances of your
winning?

Suppose, instead of losing at the second stage, I give you a third and
final chance, and if you roll a 6 you win, otherwise you have lost.
Draw a tree to represent this. What are the chances of your winning
now?

3 (*Open-ended question*) A completely fair game would have a 0.5 probability of winning and a 0.5 probability of losing. For example, tossing a single coin – heads you win, tails you lose – is a completely fair game. But this involves only one action (tossing the coin), unlike our games above, which have more than one step to them. Can you come up with other games involving at least two actions (e.g. toss a coin, and then roll a die, with various combinations for winning like the above) that are completely fair?

Expected values and decision criteria

Calculating the expected value of a random variable, and criteria for making decisions

In probability, you cannot know what will happen, but you can give a measure of what is likely to happen over a long period of time. Using this information, you can then make decisions as to whether a game is worth playing, a decision worth taking, and so on.

Key topics
- Expected values
- Decision criteria

Key terms
expected value expected value criterion utility function minimax
maximax and regret criteria odds

With a probabilistic event, you can never be certain what will happen: that is the whole nature of probability. But you can give some indication of what you 'expect to happen in the long run': what will be the 'average' value of a roll of a die, say, over many rolls? We shall investigate this here.

● Expected values

Probably the best way to view an *expected value* is to think of it as 'what would be the overall effect in the long run?' For example, the game of roulette is played with a ball that spins around a wheel containing 37 segments – 18 are red, 18 are black, and 1 is green – and eventually settles into one of the segments. You can gamble at each turn whether the ball will land in a red segment or a black segment (if it lands in green, you automatically lose). If you win, you get twice your money back (so if you bet £1, you get £2 returned and so make a profit of £1). If you lose, you simply lose your bet.

Regardless of whether you choose red or black, your chances of winning are the same, namely $\frac{18}{37}$ (18 of the 37 segments are 'your' colour).

Let's suppose you bet £1 every time you play. You have a $\frac{18}{37}$ chance of winning, and so a $\frac{19}{37}$ chance of losing. Suppose you play 1000 times. How much money would you expect to come out with? Note the following facts:

- Every game is essentially the same: whether you choose red or black, you have the same probability of winning each time (18/37).
- For a winning game, probability 18/37, you get £2.
- For a losing game, probability 19/37, you get £0

Now, if you multiply the probabilities by the outcomes and then add them up, you get $2 \times \frac{18}{37} + 0 \times \frac{19}{37}$, which works out as 0.973 to 3dp. So for every game you play (which costs £1), you would expect to gain back on average £0.973, so the odds are slightly against you (that is why the casino has a green segment, to sway the probabilities slightly in their favour).

So if you play 1000 times, multiply this figure by 1000 and you get an expected return of £973. This is what you can expect to finish with. Of course, you might get lucky and win more than this, or get unlucky and lose much more, but you can expect to leave the game with a loss of around £27. If you enjoy playing the game, many gamblers would consider that a price worth paying for the entertainment value, and the casino would certainly be happy to have your money!

>
> In general, you can work out the expected value of something in this way: for each possible outcome, multiply its *value* by its *probability*, and then add up all these answers.

Let's do a few examples to illustrate this.

Example 1

What is the expected value of a roll of a standard die?

There are six different possibilities for the roll: 1, 2, 3, 4, 5 or 6. The probability of each of them is $\frac{1}{6}$. So multiplying together the values and the probabilities, and then adding them up, you get:

$$1 \times \frac{1}{6} + 2 \times \frac{1}{6} + 3 \times \frac{1}{6} + 4 \times \frac{1}{6} + 5 \times \frac{1}{6} + 6 \times \frac{1}{6} = 3.5$$

Thus, using our 'in the long run' view of the expected value, overall you can expect to average 3.5 from a roll of a die. Note that it's not actually possible to roll the expected value, so it's not 'expected' in the sense that it's the most likely to happen. You can't roll 3.5, but in the long run, the average score of your rolls is likely to be around 3.5.

Example 2

You give me £1 and then roll a die. If you get a 6, then I give you £5, otherwise I keep your money. What is your expected return?

In this example, there are two possible outcomes. With probability $\frac{1}{6}$ you roll a 6 and win £5. With probability $\frac{5}{6}$ you roll something other than a 6 and win nothing. So your expected return is:

$$5 \times \frac{1}{6} + 0 \times \frac{5}{6} = 0.833 \text{ (3dp)}$$

So you can expect to win roughly 83.3p for every £1 you spend. I don't suggest you play this game!

Example 3

Just as above, you give me £1 and then roll a die. If you get a 6, then I give you £5, otherwise I give you 20p. What is your expected return now?

In this example, there are two possible outcomes. With probability $\frac{1}{6}$ you roll a 6 and win £5. With probability $\frac{5}{6}$ you roll something other than a 6 and win £0.20. So your expected return is:

$$5 \times \frac{1}{6} + 0.2 \times \frac{5}{6} = 1$$

So you can expect to win £1 for every £1 you spend, and in the long run you will probably break even (not win or lose anything).

● Decision criteria

Often we are faced with a situation where we need to make a decision one way or another on what to do. In such a position, probability can sometimes be used to help us make that choice.

We will illustrate by a game such as the original version of the TV show *Who Wants To Be A Millionaire?*, in which contestants answer questions (choosing from four possible answers), earning themselves more and more money, climbing up a 'ladder' up to a possible £1 million. When it comes to the final question, the player will currently be on £500,000. They can choose to take that money and walk away, or they can try to answer the final question. If they get it right, they win £1,000,000. If they get it wrong, they fall a long way down the 'ladder' and take away £32,000.

Suppose you are in this situation, and you have no idea about the answer to the last question. Is it a good idea to just guess? If you get it wrong, you lose a lot of money (but still win £32,000); on the other hand if you get it right, you win £1 million. What would you do? Different people would do different things: the conservative type would stay with the £500,000, whereas other people might take a 'I came with nothing' attitude and would gamble. What would you do?

We can try to answer this mathematically. What we'll do is draw a tree similarly to before. The only difference between this tree and the previous ones is that the first branch represents a *choice*, so it doesn't have any probabilities associated with it. We can choose to go either one way or another. Essentially, you can think of this as two separate trees: one for the first option and one for the second.

If you choose to gamble, you have a $\frac{1}{4}$ probability of winning £1,000,000 (picking the right answer out of four) and a $\frac{3}{4}$ probability of winning £32,000 (picking one of the three wrong answers). If you don't gamble, you have a certain probability 1 of winning £500,000. So the tree is as shown in Figure 14.1. Should we gamble or not?

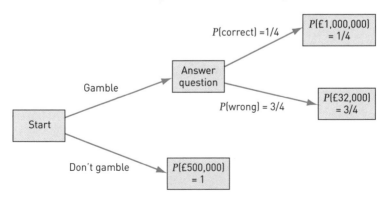

Figure 14.1 **Example of a probability tree showing decision criteria**

● Expected value criterion

One simple way to look at this problem is to work out the expected value for each choice.

If we choose not to gamble, the expected value is just £500,000: we are guaranteed to end with that.

If we do gamble, the expected value is $1,000,000 \times \frac{1}{4} + 32,000 \times \frac{3}{4}$ $= 274,000$. So the expected value if you gamble is £274,000. (Note again that this isn't a possible amount you can actually win.)

The *expected value criterion* simply says that we should pick the option that gives us the highest expected value. So we should not gamble, and should take our £500,000.

● A similar game

In this game, a player has the option of using 'lifelines' during the game. Suppose, on this last question, they still have their '50/50' lifeline. This means that instead of having four options to choose from, two wrong options are eliminated, and so they now have only two, and the probability of getting the question right is now $\frac{1}{2}$.

Would you gamble now? The tree now looks like Figure 14.2.

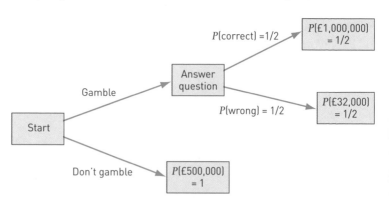

Figure 14.2 **Example of a probability tree showing decision criteria with the option of lifelines**

Now use the expected value criterion again. If you don't gamble, you are certain to leave with £500,000.

If you do gamble, the expected value is $1{,}000{,}000 \times \frac{1}{2} + 32{,}000 \times \frac{1}{2} = 516{,}000$. So, in this case, the expected value criterion would say that you should gamble.

What would you actually do? There are a couple of problems with this approach. One is that the expected value gives the 'average over a long run', but you just have one chance, and one chance only. More importantly, the amount of money involved means different things to different people. For a lot of people, £32,000 is a tidy sum but £500,000 would change their life, so the difference between them is huge, whereas the difference between £500,000 and £1,000,000, although mathematically greater, might not mean that much of a difference to someone: they might do pretty much the same with £500,000 as they would with £1,000,000, so why gamble?

If you aren't sure about this, would you gamble £1 billion on the toss of a coin, where you get £2 billion if you win and £1 if you lose? Mathematically (check this yourself) the expected value criterion says 'gamble'. But would anybody really risk £1 billion to try and get £2 billion, when if they lose, they go home with £1? Almost nobody would gamble in this situation, since they wouldn't do anything different with £2 billion than they would with £1 billion.

● Utility functions

This leads on to the idea of utility functions. Let's put ourselves back into the situation we were just talking about, and suppose we are a fairly typical working person. What we'll do is try to assign some sort of value to our winnings that indicates how much use we can actually make of them.

With £32,000, let's say we can pay off a few things and have a decent holiday. As this is the lowest outcome, assign it a value 1.

With £500,000, we might be able to buy a dream car and house, and do many things we always wanted to do. This is life-changing: let's say this has value 10, i.e. 10 times as important to us as the smaller prize.

With £1,000,000, we won't actually do much more than we did with £500,000, we'll just have more money in the bank. So this isn't that much more valuable to us than £500,000; let's say this has value 11.

The crucial point is that the difference between £32,000 and £500,000, and then £500,000 and £1,000,000, isn't much different mathematically, but in terms of what it means to us, the difference is huge.

Note of course that the choice of values is dependent on the person, and is subjective to a great extent; you would probably allocate the values differently.

Now redraw our tree as above, but this time with our utility function values rather than the actual financial amounts, so values 1, 10 and 11 rather than £32,000, £500,000 and £1,000,000 (Figure 14.3).

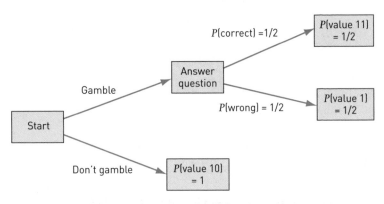

Figure 14.3 **Example of a probability tree showing 'utility function' values**

Now look at the expected values. From not gambling, you have a guaranteed value of 10. If you gamble, the expected value is

$$11 \times \frac{1}{2} + 1 \times \frac{1}{2} = 6.$$

So the best strategy now is not to gamble: it is of higher 'value' to you to take the money and not gamble.

Remember: a utility function assigns a value to an outcome in terms of how important it actually is to you. It's up to you to determine the relative values of the outcomes, depending on what they mean to you.

Go back and repeat this calculation with your own utility function. When trying to understand a concept, experiment with your own values and try to put it in your own personal context. If you were in this situation, would you personally gamble? Try to put subjects in the context of your own life, and they become much more real and relevant.

● Minimax, maximax, and regret criteria

Provided you can come up with an accurate utility function to represent your own circumstances, then the expected value approach is usually the best way to make a decision as to which option to take. But, of course, it's not the only way. We shall briefly discuss some other very basic criteria that can be used to help make a decision, which don't take the probabilities into account.

● The *minimax criterion* says that you should minimise your maximum loss, or equivalently maximise your minimum gain – essentially, be as conservative as possible. Suppose you are in the same situation again as above, on £500,000 and deciding whether to gamble or not. If you gamble and the worst happens and you lose, your gain is only £32,000. If you don't gamble, your gain is £500,000. So you should maximise this, and not gamble.

● The *maximax criterion* says that you should maximise your maximum gain – essentially, be as brave as possible. In the same situation, if you gamble and you are lucky enough to win, your gain is £1,000,000. If you don't gamble, your gain is £500,000. So you should maximise this, and gamble.

● The *regret criterion* measures how much you will regret your choice if you take the wrong option. If you gamble, and get it wrong, you will regret it, because you have lost £468,000. If you don't gamble and you would have been right, you will regret it by £500,000. So you should gamble.

These are very naive measures – they don't take into account probabilities, utility values, or anything like that – but they can be used in simple situations to form ideas as to what to do.

Now let us look briefly at what is meant by the odds that are given by bookmakers, in a horse race for example. In a perfectly even race, every competitor would have exactly the same chance of winning (creating odds like this is an example of what is known as the *Laplace criterion*, where if you do not know the probability of many events, assume they are all the same), but this is very rarely the case; after all, the competitors in a race have different skills, and some are much more likely to win than others. Instead, the bookmakers will analyse previous form, the conditions, and various other factors, and come up with their estimate of the chances of each competitor.

The odds can be interpreted as simple probabilities. Odds of 9/1 for example (read this as 'nine-to-one'; it is not a division) means that if the race was run 10 times, the competitor would be expected to lose 9 of them and win 1 of them, and so odds of 9/1 are equivalent to the probability $\frac{1}{10}$ of winning (1 out of 10 races is a win). Odds of 7/2 mean that, out of 9 races, they would expect to lose 7 and win 2, so 7/2 is equivalent to the probability of winning of $\frac{2}{9}$, and so on. In general, you can see that the odds a/b are equivalent to the probability $\frac{b}{a+b}$.

If a competitor is more likely to win than lose, they are what is known as 'odds-on'. For example, odds of 1/2 mean that they would be expected to lose 1 and win 2 if the race was run three times, so the probability of their winning is $\frac{2}{3}$.

If you win the bet, the odds represent what you win: odds of 9/1 mean that for every £1 you bet, you win £9, and you also get your stake money back. So, for example, if you bet £5 at 9/1, you get back £45 (9 × £5), plus your £5 stake, so you get £50 back from the bookmaker. Alternatively, you can view this as the probability $\frac{1}{10}$ and read off from that, that if you bet £1 you will get £10 back, so if you bet £5 you will get £50 back.

The bookmakers will create the odds so that the probabilities do not add up to exactly 1, but a little more than 1: this means that overall they would expect to make money.

● Accumulators

Accumulators are bets for which you bet on more than one event, and have to get all the bets right to win. Since this is an 'and' question (you need the first event to win, and the second, etc.), you need to multiply the probabilities. For example, if you bet on three football teams to win at odds of 2/1, 3/1 and 4/1 as an accumulator, then the probability of all three winning is $\frac{1}{3} \times \frac{1}{4} \times \frac{1}{5} = \frac{1}{60}$, and so if you bet £1 and all three teams win, you will get £60 back from the bookmaker.

Note that in an accumulator, if any one of the bets fails, you lose completely, so if two of your three teams win, but one loses, then you don't get anything back. It is possible to make 'permutation' bets where you will still win something if two out of the three win, but it is all still just probabilities.

The whole gambling industry really is just probability. The bookmakers make their probabilities; if the gambler thinks they have got it wrong, they might gamble on it. This is a good, practical example of the topic. Similarly, much of finance and banking is probability and statistics; choosing whether to buy or sell depends on your analysis of the situation and the likely probability of success.

 ## Summary

The expected value is very useful, as it gives an indication as to what you can expect to happen 'over a long term'. But you have to remember that it cannot predict exactly what will happen. Also, remember that the expected value might not be a possible value to obtain (such as the expected value of the roll of a die being 3.5), and so you can't interpret it as what *will* happen; perhaps it is better thought of as 'What is the probable average of what will happen if the action is repeated many times?'

 ## Exercises

1 A standard six-sided die has the values on its faces changed. What is the expected value of a roll of the changed die?

Example: Faces labelled 1, 1, 4, 4, 4, 4

Solution: The probability of getting 1 is $\frac{2}{6} = \frac{1}{3}$, and the probability of getting 4 is $\frac{4}{6} = \frac{2}{3}$. Hence the expected value is $1 \times \frac{1}{3} + 4 \times \frac{2}{3} = \frac{1}{3} + \frac{8}{3} = \frac{9}{3} = 3$.

(a) 1, 1, 1, 6, 6, 6,

(b) 1, 1, 3, 3, 6, 6

(c) 2, 3, 4, 5, 6, 7

(d) 10, 20, 30, 40, 50, 60

(e) 4, 4, 4, 4, 4, 4

2 A game costs £5 to play. Once you have paid, you toss two coins. If you get two heads, you get £8 back. If you get two tails, you get £11 back. Otherwise, if you get a head and a tail, you win nothing. What is the expected value of your winnings when playing this game? Can you make a small change to the rules of this game to make it completely fair (so that your expected winnings are £5)?

3 On a game show, you have to choose from five boxes. The boxes contain £1, £10, £100, £1000, and £10,000, but you don't know which is in which box. You choose a box, and it contains £1000. You are offered the chance to swap it for another box. Using each of the following criteria, should you swap?

● minimax criterion

● maximax criterion

● regret criterion

● expected value criterion

You need £800 urgently to pay off a debt before the bailiffs come round. Using the utility function below, should you swap?

Points	Utility value
1	1
10	2
100	6
1000	100
10,000	110

4 How much do you get back from the bookmaker if you win the following bets?

Example: £5 at odds of 6/1

Solution: 6/1 is the same as the probability $\frac{1}{7}$: for every £1, you get back £7. So you get $7 \times £5 = £35$. (Alternatively, if the odds are 6/1, then you get back $6 \times £5 = £30$, plus your stake £5, and so you get £30 + £5 = £35.)

(a) £2 at odds of 4/1
(b) £10 at odds of 13/2
(c) 50p at odds of 100/1
(d) £1 on an accumulator of two bets at 3/1 and 4/1
(e) £5 on an accumulator of three bets at 10/1, 12/1 and 16/1

5 In 1996, Frankie Dettori won all seven races at Ascot, with odds on the seven horses of 2/1, 12/1, 10/3, 7/1, 7/4, 5/4 and 2/1. How much would you have won if you placed a £10 accumulator bet on this?

Calculating the probability of something happening when you already have some information

Sometimes you need to work out the probability of something happening 'conditional' on another event: so you need to work out probabilities when you already have some information.

Key topics

● Conditional probability

Key terms

conditional probability Bayes' theorem

The chances of your winning the lottery at the start of the draw are highly unlikely; you need all six numbers to come up. But suppose you are watching, and already know your first five numbers have come up. Now your chance is much higher, as you just need to match the last number to win. This is an example of a *conditional probability*.

● Illustration of conditional probability

Suppose you toss two coins. What is the probability that you will get two heads? We did this in Chapter 11: the probability is $\frac{1}{4}$.

But suppose that you have already looked at the first coin, and it has come up heads. Now what is the probability that you will have two heads?

Already knowing that the first coin is heads, all you are now relying on is that the second coin has also come up heads: this is just the probability that the second coin is heads, which is now simply $\frac{1}{2}$.

The prior knowledge obviously changes the probability. As another example, if you know the first coin was tails, you already know you have no chance of getting two heads, regardless of what the second coin is.

It seems obvious, but the more you know about something, the better you can deal with it. Get all the information you can about the problem you are dealing with, so that you can make a more informed decision. No-one would gamble on two heads if they already knew one coin was tails, would they? Do the same in your studies. Find out all you can about your problem, so that you can have the most likely chance of making the best choice.

● The conditional probability formula

The notation $P(A|B)$ is used for conditional probability. This notation means the probability of event A, given that event B has happened. As we often do, let's give the formula first, and then illustrate it with an example.

$$P(A|B) = \frac{P(A \cap B)}{P(B)}$$

This formula says that the probability of something (A) happening, given that something else (B) has already happened, can be worked out by dividing the probability of both things A and B happening, by the probability of B happening.

Read this sentence again, looking at the formula, do the example that follows, and then come back and read this formula and sentence again. Often things won't click the first time, so come back to them when you have done an example and see if it makes it clearer.

Let's suppose that we have a group of 12 students. There are 8 men (3 of whom are married) and 4 women (3 of whom are married), and a class representative has been chosen randomly from this group. Let us illustrate the formula with a couple of examples:

Example 1

We know a man was chosen. What is the probability that he is married?

This is easy to solve directly: of the 8 men, 3 are married, and so the probability is $\frac{3}{8}$. We can check that the formula seems to match this. We shall use A for the event 'the person is married' and B for the event 'the person is a man'.

- $P(A|B)$ means 'the probability of A given B', or in our case 'the probability the person is married, given that they are a man'. We calculated this directly as $\frac{3}{8}$, so we are hoping the formula will give us this answer.

- $P(A \cap B)$ is the probability that the person chosen was both a man and married. Out of the 12 people, there are 3 married men, and so the probability is $\frac{3}{12} = \frac{1}{4}$.

- $P(B)$ is the probability that the person is a man. Out of the 12 people, 8 are men, and so the probability is $\frac{8}{12} = \frac{2}{3}$.

Now check the formula:

$$\frac{P(A \cap B)}{P(B)} = \frac{^{1}/_{4}}{^{2}/_{3}}$$

This takes a bit of fractional manipulation (it is $\frac{1}{4} \div \frac{2}{3} = \frac{1}{4} \times \frac{3}{2} = \frac{3}{8}$ using the usual rules of fractions, or just do it on a calculator) but you do get the same answer: the formula works!

> ✔ You can't 'prove' that a formula works all the time just with one example, but it does help to reassure you. If you create your own formulae in your studies, then test them with a few examples. If a formula seems to work, then you can be pretty confident you got it right.

Example 2

We know a married person was chosen. What is the probability that they are a man?

This is in a sense the 'opposite' question to before: now we are asking for 'the probability of being a man, given that they are married', which is the other way round. So, using A and B as before, what we want now is $P(B|A)$. The expression for this is virtually the same as the one above:

$$P(B|A) = \frac{P(A \cap B)}{P(A)}$$

Let's check the formula with this:

- Just as above, $P(A \cap B)$ is the probability that the person chosen was both a man and married. Out of the 12 people, there are 3 married men, and so the probability is $\frac{3}{12} = \frac{1}{4}$.
- $P(A)$ is the probability that the person is married. There are 6 married people in total (3 men and 3 women), and so the probability of being married is $\frac{6}{12} = \frac{1}{2}$.

Hence, using the formula, we have

$$P(A|B) = \frac{P(A \cap B)}{P(B)} = \frac{1/4}{1/2}$$

which works out to be $\frac{1}{2}$. This makes sense: there are 3 married men and 3 married women, so if you know the person is married, the probability of their being a man is $\frac{3}{6} = \frac{1}{2}$.

So, the formula seems to be working correctly.

● Bayes' theorem

You can combine the two formula above (for $P(A|B)$ and $P(B|A)$) together (try to do this for yourself as an exercise) to get what is known as *Bayes' theorem*, which links together $P(A|B)$ and $P(B|A)$.

Bayes' theorem: $P(A|B) = \dfrac{P(B|A) \times P(A)}{P(B)}$

In our examples above, we had $P(A|B) = \frac{3}{8}$, $P(B|A) = \frac{1}{2}$, $P(A) = \frac{1}{2}$ and $P(B) = \frac{2}{3}$. You should check for yourself that these values give the right answer.

● A more complex example of conditional probability

Topics such as conditional probability crop up often; a good example is in medicine. The following is a well-known example that is commonly used to illustrate this.

Example

Suppose that a test for a disease is known to be 99% reliable (so that it gives the correct diagnosis in 99% of cases), and only 1% of the population has the disease. If you are selected at random from the population to take the test, and test positive for the disease, how much should you worry?

Given the high reliability rate of the test, it seems natural that you would be very worried! Let's do the maths. This is a conditional probability; you need to work out the probability of your actually having the disease, given that you tested positive.

Let A be the event 'you have the disease' and B be the event 'you tested positive'. Remember that

$$P(A|B) = \frac{P(A \cap B)}{P(B)}$$

$P(A \cap B)$ is fairly easy to work out. This is the probability that you have the disease, and tested positive. There is only a $\frac{1}{100}$ probability of having the disease, and if you have, then the test is 99% accurate, so the probability of testing positive is $\frac{99}{100}$, so the probability of your having the disease and testing positive is $\frac{1}{100} \times \frac{99}{100} = \frac{99}{10,000}$

$P(B)$ is slightly trickier: this is the possibility of a positive result of the test. There are two options for a positive test: either you do have the disease (as we worked out, this has probability $\frac{99}{10,000}$), or you do not have the disease but still test positive. The probability of your not having the disease is $\frac{99}{100}$, and the probability of your testing positive is then $\frac{1}{100}$, so the probability of not having the disease and testing positive is $\frac{99}{100} \times \frac{1}{100} = \frac{99}{10,000}$. Since these possibilities are mutually exclusive, we can just add them up, to get that the overall probability of a positive result is $\frac{99}{10,000} + \frac{99}{10,000} = \frac{198}{10,000}$.

Hence, finally, to work out P(have the disease|tested positive) you work out $\frac{99/10,000}{198/10,000}$, which works out to be $\frac{1}{2}$.

This means that out of all the people who have tested positive, only half of them actually have the disease; so is this test a good thing? Half of the people it diagnoses as positive for the disease will go through the stress of believing they have the disease, and possibly suffer side-effects from the treatment, when they weren't even ill in the first place.

This doesn't mean the test is useless, as tests are rarely carried out on random people. The doctors would probably only do the test if you had other symptoms of the disease as well, but it's an interesting point as to how things aren't always what they seem.

> Be careful how you interpret what you are told. The manufacturers of this test would no doubt claim that their test is fantastic (99% accurate) and hospitals should rush out and buy it, but half the time it comes up positive, the patient doesn't actually have the disease. By increasing your knowledge of statistics, you understand more about claims that are made, and can understand what they actually mean.

Note, by the way, that this is a good example to illustrate the difference between $P(A|B)$ and $P(B|A)$. $P(A|B)$ is the probability of having the disease, given that you tested positive, which we calculated to be only $\frac{1}{2}$ (or 50%), whereas $P(B|A)$ is the probability of testing positive given that you have the disease, which is 99% (the test is 99% reliable), or a probability of $\frac{99}{100}$. I hope you can see that these are very different things!

 ## Summary

Conditional probability is a difficult topic, but an essential one. Once you have knowledge about something happening, then you can make further conclusions; conditional probability allows you to do this. Remember, it's all about 'what can I conclude, now I've got this additional information?'

 ## Exercises

You may be able to answer these questions directly, but try to use the formula as discussed, at least to check that it works.

1 As in the examples in the chapter, calculate:

(a) The probability that a married person was chosen, given that it was a woman.

(b) The probability that a woman was chosen, given that it was a married person.

2 As in the question in Chapter 12, a bag of children's toys contains 3 red circles, 2 green triangles, 2 green squares, 2 red squares and 1 red ball, from which a child selects one item at random. Calculate the following:

Example: The object is a square, given that it is red.

Solution: Let A be the event that 'it is square', and B be the event that 'it is red'. Then the probability we are looking for is $P(A|B) = \dfrac{P(A \cap B)}{P(B)}$. $P(A \cap B)$ = P(square and red), which since there are 10 toys and 2 red squares, is $\dfrac{2}{10} = \dfrac{1}{5}$. $P(B) = P(\text{red}) = \dfrac{6}{10} = \dfrac{3}{5}$. Then $P(A|B) = \dfrac{P(A \cap B)}{P(B)} = \dfrac{1/5}{3/5} = \dfrac{1}{3}$. (You may have been able to see this directly: there are only 6 red objects in total, 2 of which are square, so the probability is $\dfrac{2}{6} = \dfrac{1}{3}$.)

(a) The object is a circle, given that it is red.
(b) The object is green, given that it is a square.
(c) The object is green, given that it is a ball.

PROBABILITY DISTRIBUTIONS

Introduction to probability distributions

An introduction to the idea of probability distributions, and the differences between them

A random event may not be spread equally among the various possibilities; some may be more likely than others. Here we shall investigate how the various possibilities may be spread, before moving on to investigate specific distributions in the next few chapters.

Key topics
- Probability distributions
- Averages and spread of probability distributions
- Skewness and kurtosis

Key terms
probability distribution discrete continuous mean median mode variance standard deviation skewness kurtosis

If you roll a die, then every possibility (1, 2, 3, 4, 5, 6) is equally likely; you can say that the probabilities are *uniformly distributed*. But for most things this isn't the case.

Suppose you roll two dice and add the values together. It's possible to get any value from 2 to 12, but some values are more likely than others. For example, you can get 2 in only one way (a 1 and a 1 on the two dice), but there are lots of ways to get 7 (for example a 1 and 6, or 6 and 1, or 2 and 5, and more), so 7 is more likely than 2.

This is an example of a probability distribution, which essentially records the likelihood of each possibility occurring. We shall discuss the idea of probability distributions fairly informally, and in general, in this section, before we move on to look at specific distributions over the next few chapters.

Probably the best way is to introduce this concept by example; we shall look at two examples here to illustrate the idea of probability distributions.

● Discrete probability distributions

Recall the idea of the word *discrete*: this means that the possible values can be one of a fixed number of distinct possibilities. Consider for example, as suggested above, the total value of two dice. This value could take several possible values – in fact any whole number between 2 and 12. It can't take any other values (and nothing in between, there are no fractions, etc.), so this is indeed a set of discrete values.

However, some values are more likely than others. As we suggested above, 7 is more likely than 2, because it can be made in many different ways.

How many possibilities are there for the roll of two dice? The first die can take one of six values, and the second die can take one of six values, so there are $6 \times 6 = 36$ possibilities.

These 36 possibilities are (1, 1), (1, 2), (1, 3), (1, 4), (1, 5), (1, 6), (2, 1), (2, 2), ... and so on (where each pair comprises the roll of the first die and the roll of the second die). Each of these has a total: so for example the pair (3, 4) has total 7, the pair (5, 6) has total 11, and so on.

 Exercise

Write down these 36 possibilities, and for each one write down the total you get (you could do this as a 6×6 grid if you find it easier!)

Now write down how many ways there are to get each value from 2 to 12. For example, there is only one way to get a total of 2 (the pair (1, 1)) but there are two ways to get 3 (the pairs of (1, 2) and (2, 1)), and so on. If you count right, you should get the following numbers of ways. Note that, since there are 36 possibilities, the probabilities in the last column are simply the number of ways divided by 36.

Total	No. of ways	Probability
2	1	1/36
3	2	2/36 = 1/18
4	3	3/36 = 1/12
5	4	4/36 = 1/9
6	5	5/36
7	6	6/36 = 1/6
8	5	5/36
9	4	4/36 = 1/9
10	3	3/36 = 1/12
11	2	2/36 = 1/18
12	1	1/36

You can see from this that the 'middle' numbers are more likely and the 'outside' numbers are less likely: this is very common with probability distributions.

Don't just *read* books and lecture notes; *do* them (such as the exercise above). It's all too easy to read something and think 'I get that, I won't bother doing it', but when you come to do it in an exam, you realise that actually you *don't* know how. Statistics is one of those subjects that is far better done by practice than by reading and just thinking 'I can do that' – so get your pen and paper and actually do everything. You gain so much confidence by actually doing something!

This is an example of a *probability distribution*: it explains how the various possibilities are distributed according to their probability. As with much of statistics, it is far easier to understand when you draw a picture of it. The bar chart in Figure 16.1 (with the values 2–12 along the horizontal axis, and the probabilities on the vertical axis) illustrates this distribution graphically. Again, draw this carefully for yourself. Choose a sensible scale for the vertical axis. Remember, these are probabilities, which means they are small numbers less than 1, so you would probably make your axis have labels 1/36, 2/36 = 1/18, etc.

This is a very nice distribution: it basically just goes up and down in a straight line. However, most things aren't this easy.

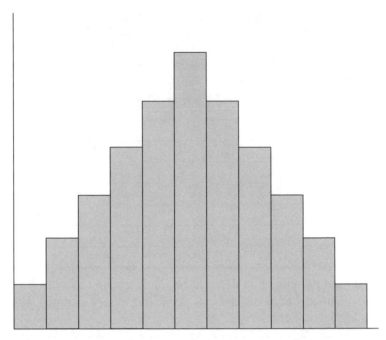

Figure 16.1 **Example of a bar chart showing probability distribution**

● Continuous probability distributions

The above is an example of a discrete probability distribution: the possible values are a finite set of fixed, distinct values. However, in many situations the distribution will be continuous; it can take any value in a range, rather than fixed, distinct values as in the discrete case. For example, you could consider the heights of people, or the speeds of cars on a road. Suppose that, by observation, you expect the probability distribution of the speed of cars to look something like the graph in Figure 16.2, where we have divided the speeds into 10 mph intervals, such as 20–30 mph, 30–40 mph, etc.

If we take a 'tighter' division, say into 5 mph intervals (25–30 mph, 30–35 mph, etc.), the graph might look more like Figure 16.3.

If you imagine refining these intervals further, into say 1 mph intervals and then even more, you can imagine that eventually this graph will just look like a curved line – perhaps something like Figure 16.4.

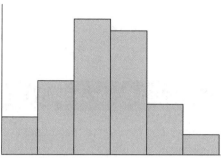

Figure 16.2 **Example of a bar chart showing probability distribution of the speed of cars divided into 10 mph intervals**

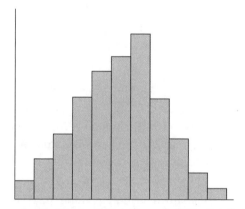

Figure 16.3 **Example of a bar chart showing probability distribution of the speed of cars divided into 5 mph intervals**

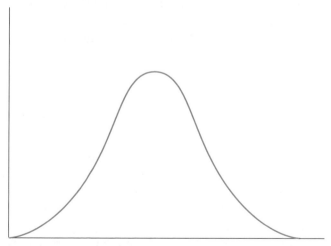

Figure 16.4 **Example of a graph showing probability distribution of the speed of cars divided into 1 mph intervals**

Introduction to probability distributions 153

When we discuss some actual continuous distributions later in this section, you will see how to use these ideas to answer quite complex statistical questions.

● Averages and spread of probability distributions

Note that this concept is very similar to the concept of frequency distributions from earlier. The only real difference is that, with a frequency distribution, we are talking about actual data we have already obtained, whereas with a probability distribution we are talking about the values of data that we expect to obtain.

We can define concepts similar to the measures of average and spread that we used before, but we need to couch them in probability terms.

- The **mean** of a probability distribution is the expected value, so it is essentially the 'average'.
- The **median** of a probability distribution is the point where a random value is less than the median with probability 1/2, and greater than the median with probability 1/2 – so the median is essentially the 'middle' value, as before.
- The **mode** of a probability distribution is the 'peak' of the distribution – the most likely value.

We can also define quartiles and percentiles similarly to the way we define the median. So, for example, the 20th percentile is the point where a random value is less than it with 20% probability, and greater than it with 80% probability.

These measures (mean, median and mode) are not too hard to work out for discrete distributions, but they are difficult for continuous distributions: these require topics such as integration, which are beyond what I want to cover in this book.

You can also define the variance similarly to before, although again this is beyond what I want to cover in this book. Recall, though, that for the variance the basic idea was to take every value away from the mean, square the values, and then divide by n, so essentially working out the 'average of the distances squared'. In a probability distribution, this can be interpreted as the 'expected value of the distances squared', and so we can define the variance in this way.

The standard deviation is just the square root of the variance, as before.

● Skewness

Most probability distributions are 'centred' around the middle. However, if a distribution has a particular tendency towards one side then it is referred to as *skewed*.

The two diagrams in Figure 16.5 show examples of skewed distributions. In the first diagram, most of the possible values are to the right of the peak: it has an early peak, and then a long 'tail'. So this distribution is said to be *skewed to the right* or to have *positive skew*. Note that 'skewed to the right' means that the peak is to the left. This can be confusing at first, but think about the fact that an early peak means that most of the possible values are to the right, and it *does* make sense.

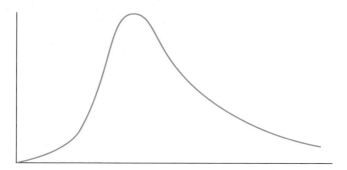

Figure 16.5 **(a) Example of a skewed distribution (skewed to the right)**

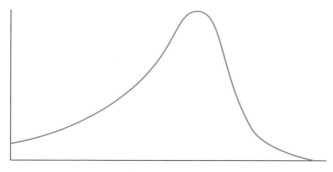

Figure 16.5 **(b) Example of a skewed distribution (skewed to the left)**

Similarly, the second diagram is *skewed to the left*, or has *negative skew*.

There are many measure of skewness that can be used. One of the simplest is *Pearson's coefficient of skewness*, which is defined as

$$\frac{3(mean - median)}{standard\ deviation}$$

That is, subtract the median from the mean, multiply by 3, and divide by the standard deviation.

● Kurtosis

As well as skewness, which measures how far to the left or right the distribution is skewed, you can also consider how 'peaked' or 'flat' it is. The name for this is *kurtosis*. For example, consider the two distributions in Figure 16.6.

The first distribution has a very pronounced peak, and so would be said to have *high kurtosis*, whereas the second distribution is much flatter, and so would be said to have *low kurtosis*.

Again, there are various ways to measure kurtosis; these are beyond what I want to discuss here, but it's useful for you to know the terms.

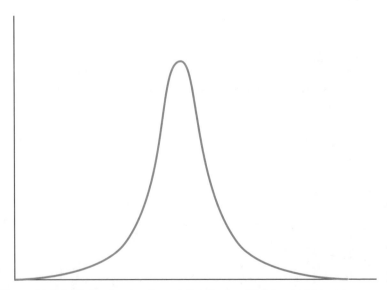

Figure 16.6 **(a) Example of a skewed distribution (high kurtosis)**

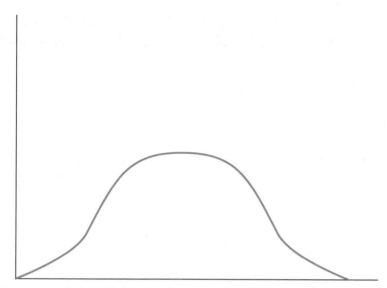

Figure 16.6 (b) Example of a skewed distribution (low kurtosis)

 Summary

You cannot predict what will happen when an event takes place, but you can know that some outcomes are more likely than others. When rolling two dice, you know that 7 is a more likely outcome than 2. Similarly, when selecting someone at random, their height is more likely to be around the average than they are to be extremely tall, and so on. The way in which the different probabilities are spread around the possible options forms a probability distribution, and allows us to analyse mathematically the likelihood of outcomes actually occurring.

As in the chapter, create a table of the number of ways to obtain the outcomes in the following questions. Then draw a graph of the probability distributions, and say whether you think they have any particular skewness and/or kurtosis.

Example: Choosing a number from the list 1, 2, 2, 3, 3, 4, 4, 4, 4, 5, 5

Solution: Drawing a table of the possible ways to choose each number, there is only one 1, there are two 2s, two 3s, four 4s and two 5s, so we get the table below.

Number	Possibilities
1	1
2	2
3	2
4	4
5	2

A graph of this looks like this:

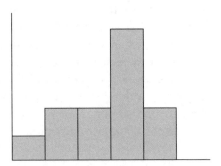

This graph is skewed to the left (more of the possible values are to the left of the peak) and has a reasonable level of kurtosis (a relatively high peak).

1 Choosing a ball at random from a bag of 20 balls, which have numbers
1, 2, 2, 3, 3, 3, 3, 3, 3, 3, 3, 3, 3, 3, 3, 3, 3, 3, 3, 4, 4, 5 on them.

2 Choosing a random number from the set
{1, 2, 3, 3, 4, 4, 5, 5, 5, 6, 6, 6, 6, 7, 7, 7, 7, 8, 8, 8, 9, 9, 10, 10}

3 Choosing a random number from the set {1, 2, 3, 4, 5}, choosing a random number from the set {2, 3, 5, 7, 11} (the first five primes), and then adding these two numbers together.

The Poisson distribution

A first example of a common discrete probability distribution

What happens when you have already observed the expected number of events over a time period? What is the probability of a certain number of events happening over another similar time period? This is the idea behind the Poisson distribution.

Key topics
● Poisson distribution

Key terms
Poisson distribution lambda e

The Poisson distribution, which was introduced by the French mathematician Siméon Poisson in the early 19th century, is our first example of a probability distribution.

Note that 'Poisson' is pronounced something like 'pwasson': it's actually the French word for 'fish', as well as the name.

✔ Make sure you know how to pronounce names and words. If you aren't sure, consult a dictionary or look online; you don't want to embarrass yourself by saying something completely wrong in front of people!

● Motivation behind the Poisson distribution

A commonly used example to illustrate the Poisson distribution is a situation like the number of calls to a company's call centre. Let's say that, from recent experience, the company knows that during working hours, they receive on average 45 calls every hour.

They would obviously like to know the probability of receiving a certain number of calls in a particular time period. The average may be 45, but how likely are they to receive 60 in a given hour? Or just 20? This sort of knowledge can play an important role in deciding how many staff to employ, and thus forms part of the company's overall strategy.

> Remember that statistics is used in industry, and is vital for planning. In this example, knowing the probability of getting more calls than you can cope with is very important when deciding how many staff to employ. Always keep in mind that these concepts are actually being used; you aren't just learning them for no reason!

The calls are assumed to be independent of each other, and are totally random events. For example, if you don't get a call for 5 minutes, that doesn't mean you are bound to get a call soon. The chance of someone calling you is the same, whether you've had no calls for 5 minutes or 15 calls in that time. A person calling doesn't know how many calls you have had; they are just calling you with their query!

This idea is important. It's the same with many things: if you are tossing coins and you've had eight heads in a row, many people think it's more likely to be a tail next, but that's just not true. You simply toss the coin, and it has an equal chance of coming up heads or tails; the coin doesn't have a brain to remember how it landed the last few times!

This gives us the basis for the idea behind the Poisson distribution. Now I'll give you the formula and explain what it means, and then we'll work through some examples.

● The Poisson distribution

The formal definition of the Poisson distribution is as follows.

Suppose an event happens on average λ times in a time period. Then the Poisson distribution says that the probability of it happening r times in a similar time period is given by the formula

$$P(r) = \frac{\lambda^r \, e^{-\lambda}}{r!}$$

A couple of things to note about the symbols used in this formula:

- The letter λ is the Greek letter lambda, and is used almost universally for the Poisson distribution, so you need to know it. The Greek alphabet is given as an appendix.

- If you haven't seen it before, e is a special mathematical number, a bit like pi (π). Written out it goes on for ever, but it is approximately 2.718281828, and is absolutely fundamental in a huge variety of mathematical and statistical areas. Your calculator should have a button for it, but use an approximation if you need to.

● Examples of the Poisson distribution

Suppose a call centre receives an average of two calls per minute. What is the probability of getting a particular number of calls? We shall assume they follow a Poisson distribution, and so the formula for r calls is $P(r) = \dfrac{\lambda^r e^{-\lambda}}{r!}$. Let us look at some example questions.

Example

What is the probability of getting exactly three calls in a given minute?

Solution

$\lambda = 2$, and we have $r = 3$ (we want three calls), and so we calculate $P(3 \text{ calls}) = \dfrac{\lambda^r e^{-\lambda}}{r!} = \dfrac{2^3 e^{-2}}{3!}$, which is approximately 0.180447.

We can use the usual rules of probability just as before.

Example

What is the probability of getting less than two calls in a given minute?

Solution

$P(\text{less than 2 calls}) = P(0 \text{ calls}) + P(1 \text{ call})$.

- $P(0 \text{ calls}) = \dfrac{2^0 e^{-2}}{0!}$ (since $\lambda = 2$ and $r = 0$), which works out to be approximately 0.135335.

- $P(1 \text{ call}) = \dfrac{2^1 e^{-2}}{1!}$ (since $\lambda = 2$ and $r = 1$), which works out to be approximately 0.270671.

Hence P(less than 2 calls) = P(0 calls) + P(1 call), which is approximately 0.135335 + 0.270671 (the two answers obtained above) = 0.406006

So although the average number of calls is two, there is a probability of approximately 0.4 that you will actually get less than two. Note that this immediately implies that there is an approximately 0.6 probability of getting two or more calls.

Now suppose we currently have enough staff to cope with up to five calls a minute. What is the probability of getting more than this in a given minute?

Example
What is the probability of getting more than five calls in a given minute?

Solution
We can't add up all the probabilities more than 5, because this could go on for ever: P(6) + P(7) + P(8) + ... But we can use complements:

P(more than 5 calls) = 1 − P(5 calls or less) = 1 − [P(0) + P(1) + P(2) + P(3) + P(4) + P(5)]. Calculating each of these to 6 decimal places:

- $P(0) = \dfrac{2^0\,e^{-2}}{0!} = 0.135335$
- $P(1) = \dfrac{2^1\,e^{-2}}{1!} = 0.270671$
- $P(2) = \dfrac{2^2\,e^{-2}}{2!} = 0.270671$
- $P(3) = \dfrac{2^3\,e^{-2}}{3!} = 0.180447$
- $P(4) = \dfrac{2^4\,e^{-2}}{4!} = 0.090224$
- $P(5) = \dfrac{2^5\,e^{-2}}{5!} = 0.036089$

Using these approximated figures,

P(more than 5 calls) = 1 − [P(0) + P(1) + P(2) + P(3) + P(4) + P(5)]
= 0.016563

So there is only approximately a 1–2% chance of getting more than five calls in a minute, but when you consider there are 60 minutes in an hour, then actually this is quite likely to happen several times a day. Perhaps you *do* need more staff?

Note that if the time period is different from the one given, you have to adjust the value of λ. For example, if there is an average of two calls in a minute, then there is an average of 10 calls in 5 minutes, 20 calls in 10 minutes, and so on.

Example

What is the probability of getting exactly five calls in 5 minutes?

Solution

We need to change the value of λ: if there is an average of two calls in one minute, then there should be an average of 10 calls in 5 minutes, so we have $\lambda = 10$.

Then $P(5 \text{ calls}) = \dfrac{10^5 e^{-10}}{5!} = 0.037833$ to 6 decimal places.

> ✔ Always be careful that you answer the question asked. The above question said 'exactly 5 calls', whereas if it said 'at least 5 calls' you would have to do a lengthier calculation, as in our third example. When it comes to an exam, or solving any problem, make sure you read the question carefully, and are certain what is being asked.

● Properties of the Poisson distribution

It's always worth knowing some useful facts about the different distributions you come across. Here are some facts about the Poisson distribution:

- The mean of the Poisson distribution is simply λ. This would be expected, since λ is essentially being defined as the average!

- Interestingly, the variance of the Poisson distribution is also λ, and so its standard deviation is $\sqrt{\lambda}$.

- The Poisson distribution is almost always positively skewed (to the right). In the call centre example, the values 0, 1, 2 go up to the mean, but then 3, 4, 5, 6, 7, 8, ... are after the mean, and so the distribution is skewed to the right. Even if the mean is very high, there are still more values after it – potentially going right up to the number of people in the world who *might* all suddenly call you at the same time!

Summary

The Poisson distribution is our first example of a discrete probability distribution. It can be a powerful tool in predicting what might happen; we saw how it could be used with call centres, in anticipating the volume of calls and the number of staff needed. Also, it is reasonably easy to use, and so quite often it is assumed that discrete events like the ideas we have discussed are following a Poisson distribution.

Exercises

1 An email account receives on average 5 spam emails a day, but the number follows a Poisson distribution. Calculate the probability of the following.

Example: The account receives exactly 3 spam emails.

Solution: $\lambda = 5$ (average per day is 5), and we have $r = 3$ (we want to know the probability of 3 spam emails), and so we calculate

$P(\text{3 spam emails}) = \dfrac{\lambda^r e^{-\lambda}}{r!} = \dfrac{5^3 e^{-5}}{3!}$, which is approximately 0.140374 to 6 decimal places.

(a) Exactly 4 spam emails are received in a day.

(b) Exactly 5 spam emails are received in a day.

(c) No spam emails are received in day.

(d) Either 4, 5 or 6 spam emails are received in a day.

(e) Less than 2 spam emails are received in a day.

(f) More than 3 spam emails are received in a day.

(g) Exactly 10 spam emails are received in two days.

(h) No spam emails are received in one hour.

2 An electricity monitoring system expects an average of 2 standard readings per hour. If it receives 1, 2 or 3 readings then that is OK, but if it receives no readings, or more than 4, it sounds an alarm. If the number of readings follows a Poisson distribution, how many alarms would you expect in a day?

(Hint: Calculate the probability that an alarm will sound in 1 hour, and hence work out how many you would expect in 24 hours.)

18 | The normal distribution

The most widely used example of a continuous probability distribution

The normal distribution is probably the most important distribution when it comes to continuous data. Relying on the fact that, in general, 'most of the values are fairly close to the average', it fits naturally into very many sets of data.

Key topics

● Normal distribution

Key terms

continuous distributions normal distribution $N(0, 1)$ $N(\mu, \sigma^2)$ distribution tables

---

The previous distribution we talked about was an example of a discrete distribution: the possible values are fixed values (usually integers). But remember that we also have continuous values, and so we need continuous probability distributions as well. The normal distribution is by far the most well known and widely used continuous probability distribution.

● An important point regarding continuous distributions

Recall that, with a discrete distribution, the outcomes can take fixed values such as 0, 1, 2, 3, etc.: there can be nothing in between. Examples of this were the roll of a die, or the number of calls to a call centre: you can't roll 2.5, and what does 'two and a half' calls mean? You either have two calls or three calls; you can't make 'two and a half calls'.

In a continuous distribution, the values can take any value, and have to be classified in ranges. For example, if a person's height is 1.62 m, it might actually be 1.6201 m, or 1.61789 m, so by '1.62 m' we really mean '1.62 m to the nearest cm'. It makes little sense to talk about the probability of a person being of 1.62 cm exactly, since it is virtually impossible that they are *exactly* 1.62 m tall: they might be 1.6200000000157... m tall! What we mean is that, to the nearest cm, they are 1.62 m tall.

Similarly, if you buy a 1 kg bag of potatoes, then some bags will be slightly smaller (say 0.99 kg), some slightly larger (say 1.01 kg), but they should all be 'acceptably' considered as 1 kg.

This means that, when we have continuous distributions, we cannot ask for the probability of a particular height: no-one will be *exactly* 1.62 m tall. But we can ask for the probability of someone being in the range ⩾1.615 m to <1.625 m, any value of which would round to 1.62 m to the nearest centimetre. So, with continuous probability distributions, we are always looking for the probability that the value lies in a range, not the probability that it takes a particular value (unlike the discrete case).

I've repeated and stressed the difference between discrete and continuous a few times in this book: that is because it is important! When you are studying a subject, it can often help to be reminded of things, so dip back into your notes and remind yourself of what you learnt before, to keep all the ideas fresh in your mind.

● Motivation of the normal distribution

If you take a sample of continuous data, you often find that 'most of' the values are reasonably close to the average, and only a few are really far away.

For example, if you consider the speed of cars on a 30 mph-limit stretch of road, some will be doing a bit more and some a bit less, but you'd expect most to be around 30. There will of course be a few outside ones – the very slow car doing 10 mph, or the speeding car doing 50 mph – but the majority of cars will be doing around 30 mph.

Similarly, if you consider the heights of people, then most people are 'around' the average, although there are some very tall and other very small people.

The idea of the normal distribution is to give a distribution reflecting this fact, which allows for most of the values to be fairly close to the mean, but also allows for a smaller number of *outliers* – values that are further away.

This is a very powerful distribution. It models a large number of situations very accurately, and is often considered as the 'default' continuous distribution.

A normal distribution is always specified by two things: its *mean* and its *standard deviation*.

It has become standard notation to represent the normal distribution as $N(\mu, \sigma^2)$, where μ represents the mean, and σ represents the standard deviation. The 'square' of the σ^2 signifies the variance: the variance is the square of the standard deviation.

Remember that the mean represents the central point, and the standard deviation represents how spread out the data is. So a normal distribution with a different mean will have a different central point, and a normal distribution with a different standard deviation will be differently 'spread out'.

As a general rule for normal distributions, you can learn the following:

- Around 68% of the values are within one standard deviation of the mean.
- Around 95% of the values are within two standard deviations of the mean.
- Around 99% of the values are within three standard deviations of the mean.

So, for example, if you have data (say the speed of cars) with mean 60 and standard deviation 5, then around 68% of the values will be between 55 and 65 (within one standard deviation of the mean), 95% of values will be between 50 and 70 (within two standard deviations of the mean), and 99% of values will be between 45 and 75 (within three standard deviations of the mean). Anything else is possible, but is highly unlikely to occur.

The normal distribution is always symmetrical about the mean (and so has no skew).

● The N(0, 1) distribution

To start with, we are going to focus on the normal distribution with mean 0 and standard deviation 1. We do this because the normal distribution with other means and standard deviations can be calculated from this one. It makes sense to study this one as our 'base point' for calculating all the other ones.

First, we'll sketch a graph of what this looks like (Figure 18.1). We're going to use this graph a lot, so get used to drawing it!

> Often, seeing things graphically makes them much clearer than just seeing complex formulae. Get used to drawing pictures and diagrams as you work out problems; humans respond well to visual things, and drawing graphs, etc. can really help you.

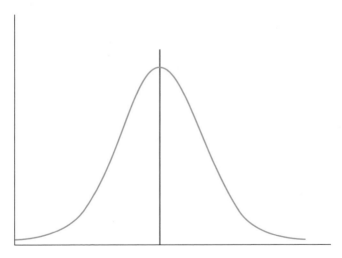

Figure 18.1 **Example of a graph showing normal distribution with mean 0 and standard deviation 1**

As you can see, the graph looks something like a bell. It has a fairly flat top, and goes wider towards the bottom. In fact this type of graph is formally referred to as *bell-shaped* (for obvious reasons).

The main point about the normal distribution is the following:

The probability of the outcome lying between two values is the area underneath this curve, between the two values.

What do we mean by this? The curve represents probabilities. The area under the whole curve can be shown to be 1, so this matches the fact that the 'total' probability should be 1.

If we want to know the probability of the value being between 1 and 2, say, then we need to work out the area of that part underneath the curve, as in the graph in Figure 18.2.

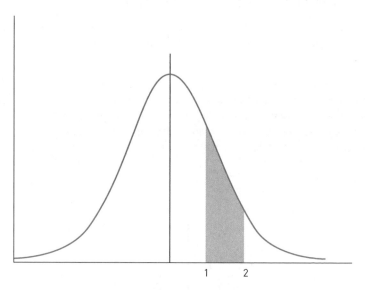

Figure 18.2 Example of a graph showing normal distribution with mean 0 and standard deviation 1 with the probability of the value within the range 1 and 2 shaded

How do we work out these areas? Sadly, there isn't a simple formula to work them out; as far as we understand, it's actually impossible to create a formula for it using our usual rules of maths.

However, there are ways to give a numerical approximation for the area. Some of them need advanced calculus and numerical analysis, and things you really don't need to know at this stage, but for now, all you need to be able to do is look up numbers in a table.

The table you need is Table A, in the Statistical Tables section at the end of this book, labelled 'The normal distribution $N(0,1)$'. What this table does is give you the area between 0 and a particular value (let's call it x); we'll see that this is enough to work out any area we need. The left-hand column of the table gives the value of x you need to look up, to one decimal place. You then look along that line to find the value you need to the second decimal place. So, for example, to look up 1.74, you go to the 1.7 row and read across to 0.04 to get the value for 1.74.

Let's give some examples of how to use this sort of table. I'm going to do this purely mathematically; we shall see more practical examples when we move into general $N(\mu, \sigma^2)$ distributions.

First, if we want to know the probability that a value is lying between 0 and some particular value (b, say), as illustrated in Figure 18.3, we can just use the table.

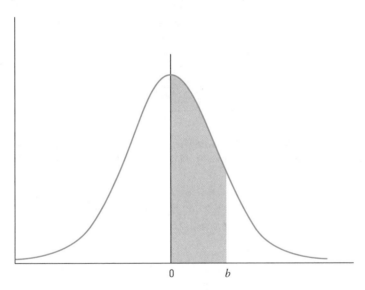

Figure 18.3 **Example of a graph showing normal distribution with mean 0 and standard deviation 1 with the probability of the value within the range 1 and b shaded**

Example 1

x is chosen randomly from the $N(0, 1)$ distribution. What is the probability that $0 < x < 1.58$ (so x lies between 0 and 1.58)?

Solution 1

The table gives us this value: if we go to 1.5 on the table and then across to 0.08, this gives us the value for 1.58; we can read off the probability as 0.4429.

If we want to work out the value between two positive points a and b, we can do this by working out the value between 0 and b, and then subtracting the value between 0 and a (which we can read off from the table). This is illustrated in Figure 18.4.

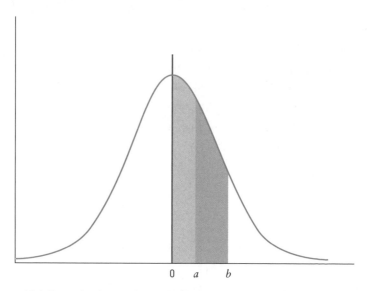

Figure 18.4 Example of a graph showing normal distribution with mean 0 and standard deviation 1 with the probability of the value within the range a and b shaded

Here the area between 0 and b is all the shaded part (both light and dark shaded). If you then take away the area between 0 and a (the light-shaded part) you are left with the area between a and b (the dark-shaded part).

Since the table gives us the area between 0 and b and the area between 0 and a, we can solve problems like this:

Example 2

x is chosen randomly from the $N(0, 1)$ distribution. What is the probability that $1.05 < x < 1.35$ (so x lies between 1.05 and 1.35)?

Solution 2

Reading off the values for 1.35 and 1.05 and subtracting them, as discussed above, we work out $0.4115 - 0.3531$, which gives the probability as 0.0584.

Suppose we want to work out the probability of a negative number lying in a range (Figure 18.5). Since the normal distribution is symmetric, then working out the area between -2.34, say, and 0 is just the same as working out the value between 0 and 2.34. So we can still use our table for negative values; essentially, we just ignore the negative sign.

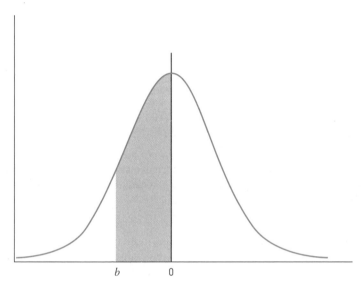

Figure 18.5 Example of a graph showing normal distribution with mean 0 and standard deviation 1 with the probability of a negative number within the range 0 and b shaded

Example 3

x is chosen randomly from the $N(0, 1)$ distribution. What is the probability that $-1 < x < 0$ (so x lies between -1 and 0)?

Solution 3

Since the value for -1 is the same as the value for $+1$, just go to 1.0 on the table and then across to 0.00 (the first entry, to give exactly 1.00); we can read off the probability as 0.3413.

Similarly, it's fairly easy to answer questions about ranges where both values are negative. As this is just a mirror image of the case where both are positive, we answer them in the same way.

The following picture (Figure 18.6) and question are almost identical to Example 2.

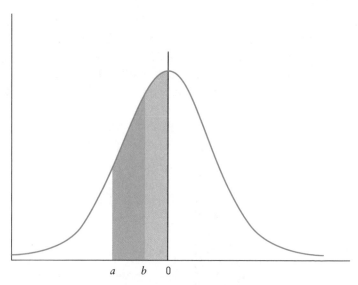

Figure 18.6 **Example of a graph showing normal distribution with mean 0 and standard deviation 1 with the probability of a negative number within the range** *a* **and** *b* **shaded**

Example 4

x is chosen randomly from the $N(0, 1)$ distribution. What is the probability that $-1.35 < x < -1.05$ (so x lies between -1.05 and -1.35)?

Solution 4

Essentially ignoring the negatives, reading off the values for 1.35 and 1.05 and subtracting them, we work out 0.4115 − 0.3531, which gives the probability as 0.0584, just as in Example 2.

Next, what happens if we want the probability of the value being between a negative number and a positive number – say between −0.5 and 1? In this case we can split the range into two: we work out the probability of the value being between −0.5 and 0, and the probability of the value being between 0 and 1 (both of which we can read from the table), and then add them together, as in Figure 18.7, where $a = -0.5$ and $b = 1$.

Example 5

x is chosen randomly from the $N(0, 1)$ distribution. What is the probability that $-0.55 < x < 1.25$ (so x lies between -0.55 and 1.25)?

Solution 5

The probability that x is between −0.55 and 0 can be read from the table

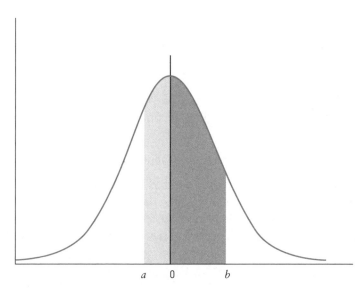

Figure 18.7 Example of a graph showing normal distribution with mean 0 and standard deviation 1 with the probability of a negative number within the range −0.5 (a) and 1 (b) shaded

(remember to ignore the negative) as 0.2088. The probability that it is between 0 and 1.25 can be read off as 0.3944, and so the total probability is 0.2088 + 0.3944, or 0.6032.

Remember that the total area under the curve is equal to 1, and so the total area under half of the curve is equal to a half, or 0.5 (Figure 18.8).

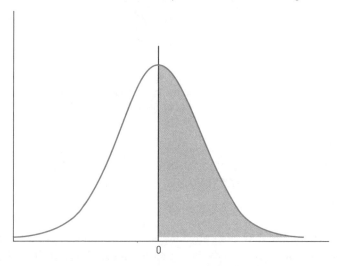

Figure 18.8 Example of a graph showing normal distribution with mean 0 and standard deviation 1 with half (0.5) of the total area shaded

Example 6

x is chosen randomly from the $N(0, 1)$ distribution. What is the probability that $x > 0$?

Solution 6

This is exactly half of the distribution, and so the answer is simply 0.5.

Of course, the probability that $x < 0$ would also be 0.5.

This can help us solve problems such as the probability that $x > 1$. In Figure 18.9, the area we need is the dark-shaded area: note that this is the whole area (which we know is 0.5), minus the light-shaded area (which we can look up from the table, as it's between 0 and 1).

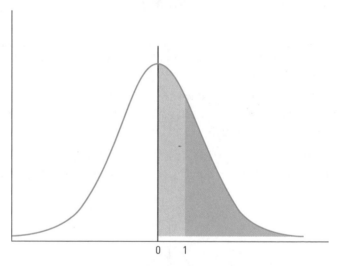

Figure 18.9 **Example of a graph showing normal distribution with mean 0 and standard deviation 1 to work out the probability that $x < 1$**

Example 7

x is chosen randomly from the $N(0, 1)$ distribution. What is the probability that $x > 1.29$?

Solution 7

The probability that x is between 0 and 1.29 can be read off from the table as 0.4015, and so the probability that x is greater than 1.29 is $0.5 - 0.4015 = 0.0985$.

As a final example, to work out the probability that $x < 1$, say, you can work out the probability of x being between 0 and 1 (from the table) and then add the 0.5 probability of x being less than 0, as illustrated in Figure 18.10, where our answer is the whole area (both the light- and dark-shaded parts).

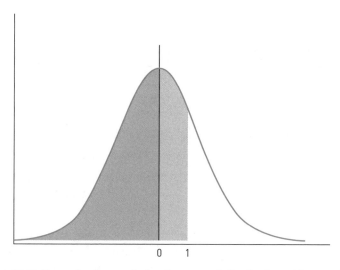

Figure 18.10 Example of a graph showing normal distribution with mean 0 and standard deviation 1 to work out the probability that $x > 1$

Example 8

x is chosen randomly from the $N(0, 1)$ distribution. What is the probability that $x < 0.75$?

Solution 8

The probability that x is between 0 and 0.75 can be read off from the table as 0.2734, and so the probability that x is less than 0.75 is 0.5 + 0.2734 = 0.7734.

This pretty much covers all the possible cases of the $N(0, 1)$ distribution. Any question you get can be answered in a similar way to the above. I *strongly recommend* that you draw a quick picture first of what you are looking for, so that you can be sure you are doing the right thing.

● Normal distributions in general

We have focused on the $N(0, 1)$ distribution because any other normal distribution (with a different mean and standard deviation) can be reduced to this, and so we can use the same table.

This allows us to investigate more practical, real-life situations that approximately follow a normal distribution. It's rare for something to follow an exact distribution, but we can say that 'it's approximately normal' to do the calculations.

Let us consider a general $N(\mu, \sigma^2)$ distribution: that is, a normal distribution with mean μ and standard deviation σ. If x is distributed like this, it is possible to show that the distribution of the variable is $z = \frac{x - \mu}{\sigma}$ just an $N(0, 1)$ distribution.

This definition makes some sort of intuitive sense. If the mean is μ and you subtract μ, it makes perfect sense that the mean will now become 0. Similarly, if you divide a standard deviation of σ by σ you should get a standard deviation of 1, so we do have an $N(0, 1)$ distribution. This is by no means a proof, but it's giving you an indication!

Don't forget the basic principle:

If x follows an $N(\mu, \sigma^2)$ distribution, then $z = \frac{x - \mu}{\sigma}$ follows an $N(0, 1)$ distribution.

As always, an example makes this much clearer. Consider the following scenario.

Scenario: The power output of a batch of 40-watt light bulbs follows approximately a normal distribution, with mean 40 and standard deviation 2.

Remember, this means that most values are close to the average 40 watts, and anything far away from 40 watts is unlikely. So you would probably be safe to market these bulbs as '40-watt light bulbs': although some are slightly less and some slightly more, there are unlikely to be any that are really far away from 40 watts.

Similar sorts of examples are buying bags of '250 g' of cat food, buying '10 mm' screws, and so on. They won't all be exactly that number, but should follow a normal distribution with (hopefully) a very small standard deviation, so that pretty much everything is acceptable. No-one is really going to mind, or even notice, if they only have a 39-watt light bulb, 249 g of cat food, or a 9.99 mm screw, and the possibility of getting something you might notice and care about (e.g. 10-watt light bulb, 150 g of cat food or a 20 mm screw) is very remote.

Note that this cannot be a true normal distribution: you can't have a negative amount of cat food or a negative length of a screw. But these possible 'negative values' are so remotely unlikely – physically impossible, in fact – that it really doesn't make any difference, and so the normal distribution is approximate enough.

Back to the example. Can we answer the following questions?

(a) What is the probability that a light bulb has a power between 38 and 42 watts?

(b) What is the probability that a light bulb has a power of more than 45 watts?

(c) What is the probability that a light bulb has a power of between 40 and 45 watts?

In this scenario, remember that the mean $\mu = 40$ and the standard deviation $\sigma = 2$, as defined.

(a) If x is the power variable, remember that $z = \dfrac{x - \mu}{\sigma}$ follows an $N(0, 1)$ distribution. So $P(38 < x < 42) = P\left(\dfrac{38 - \mu}{\sigma} < z < \dfrac{42 - \mu}{\sigma}\right)$: using the formula for z, we take away the mean and divide by the standard deviation. Since $\mu = 40$ and $\sigma = 2$, we need the probability $P\left(\dfrac{38 - 40}{2} < z < \dfrac{42 - 40}{2}\right)$, which becomes $P(-1 < z < 1)$. Since z is $N(0, 1)$ distributed, this is solved just as previously (like Example 5) from the table, and is $0.3413 + 0.3413 = 0.6826$.

(b) Similarly, $P(x > 45) = P\left(z > \dfrac{45 - 40}{2}\right) = P(z > 2.5)$, and using the techniques as before (using the table, just like Example 7), this is $0.5 - 0.4938 = 0.062$.

(c) Similarly, $P(40 < x < 45) = P\left(\dfrac{40 - 40}{2} < z < \dfrac{45 - 40}{2}\right) = P(0 < z < 2.5)$ which is immediately read off as 0.4938 from the table.

That's all there is to it. Basically, as long as you remember the formula $z = \dfrac{x - \mu}{\sigma}$, and can use the table, you can answer any question like this about normal distributions!

● A note on linear interpolation

Before we finish this section, let's make just a brief note on linear interpolation, which we first saw back in Chapter 2.

The $N(0, 1)$ table works to 2 decimal places. How would we deal with a number given to 3 decimal places, say 0.283? This should lie somewhere between the 0.28 value and the 0.29 value: using linear interpolation, it should in fact lie 3/10 of the way through the range between the values (since the next decimal place is 3).

So, since the 0.28 value is 0.1103, and the 0.29 value is 0.1141, this range is 0.1141 − 0.1103 = 0.0038. Taking 3/10 of this gives us 0.00144, so we would expect the 0.283 value to be at 0.1103 + 0.00144 = 0.11174.

This value is the value that falls in the range between 0.1103 and 0.1141 that is likely to be the closest to the extra decimal place of 3.

Linear interpolation isn't perfect (the graph isn't a straight line) but it's better than just rounding off to 0.28, and is worth doing if you need a high level of precision.

Summary

The normal distribution is one of the most important things you can learn in statistics. Many situations are modelled using a normal distribution, where most things are close to the mean, and there are only a few outliers. You'll find that many things are simply assumed to be normal, so you need to have a good grasp of this topic. Since you can reduce anything to the $N(0, 1)$ distribution, it's vital that you understand the tables and can solve problems like this; they are relevant to virtually any field of study.

Remember that probabilities have to be between 0 and 1, so if you get an answer outside this, then you know you've made a mistake. As always, it's sensible to check your answer!

 Exercises

1 Calculate the probability of the following occurring, where the variable x is chosen randomly from the $N(0, 1)$ distribution.

Suggestion: Always draw a quick sketch to isolate the area you are looking for, and see if any of the Examples 1–8 we did is similar to the problem, and if so follow the approach there.

Example: $0.25 < x < 0.75$
Solution: This is like Example 2 above. We can read off the values for 0.25 and 0.75 and subtract them, so we get a probability of $0.2734 - 0.0987 = 0.1747$.

(a) $0 < x < 0.82$ (b) $0.49 < x < 0.99$ (c) $-0.95 < x < 0$

(d) $-1.5 < x < -0.5$ (e) $-0.25 < x < 1.25$ (f) $x < 0$

(g) $x > 1.01$ (h) $x < 0.33$ (i) $x < -1$

(j) $x > -0.25$ (k) $0 < x < 0.125$ (l) $-1 < x < 0.001$

2 The amount of medicine in a bottle is approximately normally distributed, with mean 50 ml and standard deviation 0.5 ml. Calculate the following probabilities for the amount of medicine x in the bottle.

Example: $49 < x < 51$
Solution: Using the formula $z = \frac{x - \mu}{\sigma}$, we have
$$P(49 < x < 51) = P\left(\frac{49 - \mu}{\sigma} < z < \frac{51 - \mu}{\sigma}\right) = P\left(\frac{49 - 50}{0.5} < z < \frac{51 - 50}{0.5}\right),$$
which becomes $P(-2 < z < 2)$.
Since z is $N(0, 1)$ distributed this is solved just as previously (like Example 5) from the table, and is $0.4772 + 0.4772 = 0.9544$.

(a) $50 < x < 51$ (b) $50.1 < x < 50.3$ (c) $49.5 < x < 50.5$

(d) $x > 50.5$ (e) $x < 50$ (f) $x > 49.95$

3 Assume the heights of UK males follow a roughly normal distribution, with mean 177 cm and standard deviation 7 cm. Calculate the probability that a randomly selected UK male:

(a) has height between 170 cm and 180 cm
(b) is taller than 190 cm
(c) is shorter than 160 cm

The binomial distribution

A second example of a common discrete probability distribution

If you answer 10 questions in a multiple-choice test purely by guessing, how likely is it that you will get them all right and get 10/10? Or a mark of 7/10? Or a mark of 4/10? The binomial distribution allows you to answer questions like this.

Key topics
● Binomial distribution

Key terms
rare events yes/no events binomial distribution

Often we have a situation where there are only two possibilities for a particular event. For example, this could be 'right' or 'wrong' for a question, or 'head' or 'tail' for a coin toss, and so on. This leads on to the *binomial distribution*: the prefix *bi* is a common prefix referring to two, and so is an indication that there are two possibilities.

Before starting this chapter, you should go back and re-read Chapter 7 on factorials, permutations and combinations.

Don't treat every topic you study as completely separate, and think that once you have finished a topic, you don't need it any more. You need combinations for this topic, which we studied before in a previous chapter (that's why I put it before!). Keep things in your mind, and try to see your subjects as coherent wholes, rather than a mixture of unrelated things.

● Rare events and yes/no events

When we considered the Poisson distribution, we were counting events actually happening (we are receiving calls), but it doesn't really make any sense to say 'the opposite', that an event 'isn't happening', because there are billions of people on the earth who didn't call the call centre at that time – not because they tried and failed, but simply because it's absolutely nothing to do with them or what they were doing at the time.

This is known as a *rare event*: out of the 6 billion people on the earth, the likelihood of one of them calling you is pretty remote; most have never heard of you. It doesn't seem 'rare' to the call centre operatives receiving calls all day, but remember that for every person who calls, there are millions not calling, just going about their daily business.

This is different from things such as tossing a coin, where you can either succeed or fail. This is basically a *yes/no event* (Was the coin heads or not? Yes or no?). Here it *does* make sense to talk about the opposite: if the coin wasn't heads, then it was the opposite of heads, i.e. tails. With yes/no problems like this, we need a different distribution. This is where the *binomial distribution* comes in.

● Motivational example

Suppose we toss four coins. What is the probability of getting:

(a) all heads,
(b) three heads and one tail,
(c) two heads and two tails,
(d) one head and three tails,
(e) all tails?

You can solve this problem manually by listing all the possibilities. This might take you a while, so I'll do it for you: there are 16 possibilities for how the coins might fall. Using the notation H for head and T for tail, the 16 possibilities are:

(H, H, H, H), (H, H, H, T), (H, H, T, H), (H, H, T, T), (H, T, H, H), (H, T, H, T), (H, T, T, H), (H, T, T, T), (T, H, H, H), (T, H, H, T),

(T, H, T, H), (T, H, T, T), (T, T, H, H), (T, T, H, T), (T, T, T, H),
(T, T, T, T)

(I've tried to be systematic in listing these; would you be able to do this systematically yourself?)

Let's focus our attention on the number of heads, since if we know that, then we automatically know the number of tails: for example, if there is one head, there must be three tails, and so on. If you count through these 16 possibilities, this gives us the following number of possibilities for the number of heads:

Number of heads	Number of possibilities
0	1
1	4
2	6
3	4
4	1

For example, there is only one way of getting four heads (H, H, H, H) but there are many ways of getting two heads (H, H, T, T) or (H, T, H, T) and so on.

Check for yourself that this table matches with the 16 possibilities given above.

This gives us enough information to answer our problem. There are 16 possible throws, and the number of ways to make each possible number of heads is given in the table, so we can answer the questions:

(a) $P(\text{all 4 heads}) = \dfrac{1}{16}$

(b) $P(\text{3 heads and 1 tail}) = \dfrac{4}{16} = \dfrac{1}{4}$

(c) $P(\text{2 heads and 2 tails}) = \dfrac{6}{16} = \dfrac{3}{8}$

(d) $P(\text{1 head and 3 tails}) = \dfrac{4}{16} = \dfrac{1}{4}$

(e) $P(\text{all 4 tails}) = \dfrac{1}{16}$

Note that all these probabilities add up to 1 (check this for yourself), as these cover all the possible outcomes of this event.

Remember what this is all saying: if we toss four coins, we have a $\dfrac{3}{8}$

chance of getting two heads and two tails (which therefore is the most likely outcome, but still would be expected to happen less than half of the time), but we have only a $\frac{1}{16}$ chance of getting all four heads.

> ✔ Practise for yourself and convince yourself! If you have a few minutes to kill, take four coins and toss them repeatedly, noting how many heads you get at each stage. I would imagine you will get four heads or four tails rarely, but quite often get two heads and two tails. Your figures are unlikely to match the predicted values exactly, but they probably won't be far away. If you try to put things into practice and see that they do seem to work, you gain confidence and understanding that the maths and stats involved is really relevant, and does actually seem 'right'!

● The mathematics behind binomial distributions

In general, while there may be two possible outcomes, they might not have the same probability.

Suppose that the probability of 'success' (we get what we want from a particular event) is p. This means that the probability of 'failure' (we don't get what we want) must be $1 - p$, since probabilities add up to 1.

For example, suppose that there is a multiple-choice test consisting of questions that have four options. You haven't revised at all, and so are just going to guess randomly for each question. Hence the probability of getting a question right is $p = \frac{1}{4}$ (since only one of the four options can be the right answer), and the probability of getting it wrong is $1 - p = \frac{3}{4}$.

Suppose the test consists of 10 questions. What is the probability that you will score exactly 6 out of 10? This is very hard to answer as we did above by listing all the possibilities: there are so many different possibilities for the outcome of all 10 questions, which would take for ever to list, and then you would have to count how many had exactly 6 right. So let's try to think more mathematically.

The idea is to first work out the probability, given 6 specific questions from the 10, that these are all right and the other 4 are all wrong, and then work out how many possible ways there are to choose 6 questions from the 10, and multiply the probability by this amount. This should give us the overall probability. Don't panic if this isn't completely clear at first; just make sure you have some idea, and more importantly, that you can do the calculations.

So, suppose we are given any 6 of the 10 questions. What is the probability that all these 6 are correct and the other 4 are all wrong, so that we have a total score of 6/10?

- The probability that all 6 are right is $\frac{1}{4} \times \frac{1}{4} \times \frac{1}{4} \times \frac{1}{4} \times \frac{1}{4} \times \frac{1}{4} = \left(\frac{1}{4}\right)^6$.

- The probability that the other 4 are all wrong is $\frac{3}{4} \times \frac{3}{4} \times \frac{3}{4} \times \frac{3}{4} = \left(\frac{3}{4}\right)^4$.

So the probability that all 6 you chose are right, and the other 4 are all wrong, is

$$\left(\frac{1}{4}\right)^6 \times \left(\frac{3}{4}\right)^4.$$

Now, how many ways are there to choose the 6 questions? This is exactly what we did in Chapter 7: we want to choose 6 questions from 10. The order doesn't matter, since we just want them all to be right and all the rest to be wrong, so this is a perfect example of a combinations question.

Hence there are $^{10}C_6$ ways to select six questions, which is $\dfrac{10!}{6!(10-6)!}$.

Putting all of this together, there are $^{10}C_6$ ways to select six questions from the 10, and the probability of each of them all being right, and the rest all wrong, is $\left(\frac{1}{4}\right)^6 \times \left(\frac{3}{4}\right)^4$.

Hence the probability of getting 6 out of 10 in the total test is $^{10}C_6 \times \left(\frac{1}{4}\right)^6 \times \left(\frac{3}{4}\right)^4$.

If you do this on a calculator, you should get around 0.016; make sure you can get this answer. So the probability of getting exactly 6 out of 10 is around 0.016, which is pretty low.

● The binomial distribution formula

To formalise this, the binomial distribution is essentially the probability distribution that tells us the probability of getting exactly a certain number of 'successes' from a certain number of events. The most important thing you need to know is the following formula, which generalises the above idea.

The binomial distribution $B(n, p)$ says that if an event is repeated n times, and the probability of success in an event is p, then

$$P(\text{exactly } r \text{ successes from the } n \text{ events}) = {}^nC_r p^r (1 - p)^{n-r}$$

● Examples of the binomial distribution

As always, a topic becomes easier to understand when you do some examples, so here are some to try. First, we'll repeat the two examples from above using this formula, and check the answers we get there.

Note that, to work out the values of ${}^nC_r$, you can either use the nCr button on your calculator if you have one, or you can use the formula $${}^nC_r = \frac{n!}{r!(n - r)!}.$$

Example

What is the probability of getting exactly two heads if you toss four coins?

Solution

Here $p = \frac{1}{2}$ since the probability of getting a head is $\frac{1}{2}$, and we have $n = 4$ and $r = 2$ (we need two successes (heads) from four events (tosses)).

So we need to work out
${}^nC_r p^r (1 - p)^{n-r} = {}^4C_2 \left(\frac{1}{2}\right)^2 \left(1 - \frac{1}{2}\right)^{4-2} = {}^4C_2 \left(\frac{1}{2}\right)^2 \left(\frac{1}{2}\right)^2$. This works out to be $\frac{3}{8}$, or 0.375 as a decimal. This matches the answer we had before.

Example

What is the probability of getting exactly 6/10 if you have a multiple-choice test of 10 questions with 4 options for each question, and you simply guess each answer?

Solution

Here $p = \frac{1}{4}$, since the probability of getting a question right is $\frac{1}{4}$, and we have $n = 10$ and $r = 6$ (we need six successes (right answers) from 10 events (questions)).

So we need to work out

$$^nC_r p^r (1-p)^{n-r} = {}^{10}C_6 \left(\frac{1}{4}\right)^6 \left(1 - \frac{1}{4}\right)^{10-6} = {}^{10}C_6 \left(\frac{1}{4}\right)^6 \left(\frac{3}{4}\right)^4.$$ Either using

the nCr button on your calculator if you have one, or using the

formula $^nC_r = \dfrac{n!}{r!(n-r)!}$, this works out to be 0.016 as a decimal to

3 decimal places. This matches the answer we had before.

● Binomial distributions: further examples

Don't forget that we can use the normal rules of probability. For example, if we have an 'or' question, we can add probabilities, and we can also deal with complements. The following two examples illustrate this.

Example

If you toss five coins, what is the probability that you will get exactly four the same (so just one is different)?

Solution

There are two ways in which we can do this: either we can get four heads (and so one tail), or we can get one head (and so four tails). Let a head be a 'success', so $P(\text{success}) = 1/2$. Consider the two possibilities, using the usual formula:

- $P(\text{four heads}) = {}^nC_r p^r (1-p)^{n-r} = {}^5C_4 \left(\frac{1}{2}\right)^4 \left(1 - \frac{1}{2}\right)^{5-4}$, which works out to be $\frac{5}{32}$ (or 0.15625 as a decimal).
- $P(\text{one head}) = {}^nC_r p^r (1-p)^{n-r} = {}^5C_1 \left(\frac{1}{2}\right)^1 \left(1 - \frac{1}{2}\right)^{5-1}$, which works out to be $\frac{5}{32}$ (or 0.15625 as a decimal).

Hence $P(\text{four heads or one head}) = \dfrac{5}{32} + \dfrac{5}{32} = \dfrac{10}{32} = \dfrac{5}{16}$.

So the probability that exactly four of the coins are the same is $\dfrac{5}{16}$.

Often by using complements you can make a problem much easier than solving it directly, as in the following example.

Example

If you take the multiple-choice test as before (10 questions with 4 options each, where you randomly guess each question), what is the probability that you will get at least two answers correct?

Solution

Since 'at least two' means 'two or more' you could work this out by calculating the probability of exactly 2 correct, 3 correct, 4 correct, etc.: so $P(2 \text{ correct}) + P(3 \text{ correct}) + P(4 \text{ correct}) + \ldots + P(10 \text{ correct})$. This would work, but it would be a lengthy calculation, and it would be easy to make mistakes.

But it is easier to note that the complement of 'at least 2 correct' is 'either 0 or 1 correct'. Then $P(\text{at least 2 correct}) = 1 - P(\text{either 0 or 1 correct})$, by the usual rules of complements.

Remembering that $p = \dfrac{1}{4}$ (the probability of guessing the correct one of the four possible answers), to work out $P(\text{either 0 or 1 correct})$, you work out:

- $P(0 \text{ correct}) = {}^nC_r p^r (1-p)^{n-r} = {}^{10}C_0 \left(\dfrac{1}{4}\right)^4 \left(1 - \dfrac{1}{4}\right)^{10-0}$, which works out to be approximately 0.0563135 as a decimal.

- $P(1 \text{ correct}) = {}^nC_r p^r (1-p)^{n-r} = {}^{10}C_1 \left(\dfrac{1}{4}\right)^1 \left(1 - \dfrac{1}{4}\right)^{10-1}$, which works out to be approximately 0.1877117 as a decimal.

So $P(0 \text{ or } 1 \text{ correct}) = P(0 \text{ correct}) + P(1 \text{ correct})$, which, by adding up the two previous values, is approximately 0.2440252.

Hence, $P(\text{at least 2 correct}) = 1 - P(0 \text{ or } 1 \text{ correct}) = 1 - 0.2440252 = 0.7559748$.

So you have just over a 75% chance of scoring at least 2 out of 10. Of course, it's better to actually revise for the test and not resort to guessing!

● Properties of the binomial distribution

Here are some useful properties of the binomial distribution.

● The expected value (mean) of the binomial distribution $B(n, p)$ is simply np. This makes sense when you think about it: there are n events, each with probability p of success. So the expected number of heads when tossing four coins is $4 \times \frac{1}{2} = 2$, and the expected value of the score in the test of 10 randomly guessed multiple-choice questions is $10 \times \frac{1}{4} = 2.5$.

● The variance of the binomial distribution $B(n, p)$ can be shown to be $np(1 - p)$. Therefore the standard deviation is $\sqrt{np(1 - p)}$.

● If the probability $p = \frac{1}{2}$, then the binomial distribution is perfectly symmetrical (it has no skew). Otherwise, if $p < \frac{1}{2}$, then the distribution has positive skew, and if $p > \frac{1}{2}$, then the distribution has negative skew. Again, this makes perfect sense when you think about it, since if $p < \frac{1}{2}$ then we are more likely to fail than succeed, and so the peak will be early on in the small values, so we have positive skew; similarly for $p > \frac{1}{2}$ and negative skew.

● If n is large, and np is fairly small, the Poisson distribution with $\lambda = np$ is a good approximation to the binomial distribution $B(n, p)$, so it is often used instead, as it is much easier to calculate (the Poisson formula is considerably easier than the binomial formula).

 Summary

The binomial distribution is another example of a probability distribution. It is a discrete distribution (there are only a fixed, finite number of values possible, e.g. 0, 1, 2, 3, 4, 5, 6, 7, 8, 9, 10 in the test), and appears when we have a repeated series of events that can either 'succeed' or 'fail'. This is different from what we have seen before. Of course, there are many other distributions, depending on the particular situation, but the binomial distribution (along with the Poisson distribution and the normal distribution) is one of the most common.

1 8 coins are tossed. Use the binomial distribution formula
$^nC_r p^r (1 - p)^{n-r}$ to calculate the probability of the following.

Example: There are exactly 5 heads.

Solution: $n = 8$ (there are 8 coins), $r = 5$ (we want 5 heads) and $p = \dfrac{1}{2}$ (the
probability of getting a head is $\dfrac{1}{2}$), and so

$$^nC_r p^r (1 - p)^{n-r} = {}^8C_5 \left(\frac{1}{2}\right)^5 \left(1 - \frac{1}{2}\right)^{8-5} = 56 \left(\frac{1}{2}\right)^5 \left(\frac{1}{2}\right)^3 = 0.21875 \text{ (which you might}$$
give as e.g. 0.219 to 3 decimal places).

(a) There are exactly 4 heads.

(b) There are exactly 7 heads.

(c) There are no heads.

(d) There are either 4 or 5 heads.

(e) There are *not* 4 heads.

(f) There are at least 2 heads.

2 A new medical treatment that is successful 80% of the time is
used on 10 patients. Calculate the probability of the following.

Example: Exactly 7 people are cured.

Solution: $n = 10$ (there are 10 patients), $r = 7$ (we want 7 people cured) and
$p = 4.5$ (since $80\% = \dfrac{80}{100} = \dfrac{4}{5}$) and so

$$^nC_r p^r (1 - p)^{n-r} = {}^{10}C_7 \left(\frac{4}{5}\right)^7 \left(1 - \frac{4}{5}\right)^{10-7} = 120 \left(\frac{4}{5}\right)^7 \left(\frac{1}{5}\right)^3 = 0.201326592 \text{ (which you}$$
might give as e.g. 0.201 to 3 decimal places).

(a) Exactly 8 people are cured.
(b) Exactly 5 people are cured.
(c) All 10 people are cured.
(d) Either 7 or 8 people are cured.
(e) At least one person is not cured.

3 In a promotion, a fast food outlet offers a scratchcard with every
burger bought, of which 1 in 5 is a winner, and wins a free burger.
A group of 12 people comes to the outlet and decides to buy 10
burgers, in the hope that they will win at least 2 free burgers with
their scratchcards to feed everyone (and perhaps have some burgers
left over to take away for later, if they win more than 2). What is the
probability that they will be able to feed everybody in the group?

HYPOTHESIS
TESTING

Introduction to hypothesis testing

A gentle introduction to the idea of hypothesis testing

The idea of hypothesis testing is to test whether we believe something to be true by looking at a sample of data. Of course, this is only a sample, so how confident can we be that we are making the right conclusion?

Key topics
- Sampling
- Hypothesis testing
- Confidence intervals and critical values
- Errors

Key terms

sampling hypothesis testing null hypothesis alternative hypothesis Type I error Type II error confidence intervals tails one-tailed two-tailed

---

People and companies often make all sorts of claims about products: 'This is better than the old version', 'Tests show that this improves your health', 'This is proven to be better', and so on. What we shall do here is explain statistically how we might go about making such a claim, and discuss how we can be confident that we have made the correct statement.

● Sampling

Most of the time, when we want to know information about a population, we can't get every single piece of data. For example, an opinion poll is not done by asking every single person who they intend to vote for; we just wouldn't have the time and resources to do that. So we choose a small *sample* of the population, and hope that they will be representative of the population as a whole.

Obviously, the larger the sample, the more reliable it will be as an indicator of the overall data. And the smaller the standard deviation of the population, the more likely it is that the sample will be representative. If you have a large standard deviation, then the population is very spread, and there are a lot of values far from the mean, so there is more chance of the sample being unrepresentative. Compare this with a population with a small standard deviation: most values are close to the mean, and so the sample values are also likely to be close to the mean. The *standard error* of a sample of size n from a population with standard deviation σ can be defined as $\frac{\sigma}{n}$, which is a sensible measure, as it gets larger when σ is larger, and gets smaller as n gets larger, which represents both of the points made above.

A sample from a population can never be completely representative. There is always the possibility that – perhaps purely by bad luck – you choose a completely unrepresentative sample. For example, suppose you want to know the average height of the population. To do this, you sample 100 people and take the average of their heights, and hope that this is representative of the population. It may well be – but you might, by pure chance, choose 100 very tall people in your sample, and so your answer would not be close to the actual population mean. If you take a larger sample, say 1000 people, then it's more likely to be representative, but there is still the possibility that you were unlucky, and chose a lot of very tall people.

● Hypothesis testing

Hypothesis testing is all about the validity of making claims from a sample. For example, suppose a company claims that the new range of batteries they have produced lasts longer than their old range. To back up their claim, they test a sample of 50 batteries on a very high-use device, and find that the average time they last is 7.1 hours, compared with the old ones, which are known to last 7 hours on average. Is this enough evidence to support their claim?

We'll analyse this formally later, but, looking at it, how likely is it that the new batteries are genuinely better? It seems so from this sample, but because we tested only a sample (obviously we can't test them all, or we'd have none left to sell), might it be that we just got lucky and picked 50 very good batteries, and actually the average is no different from the 7 hours of the old ones?

Really, what we are asking is how *confident* we can be in the claim we are making. If we can be 99% sure, we can be very confident we are right: there is only a 1 in 100 chance that we are wrong by bad luck. If we can be 95% sure, then we can be reasonably confident, but now there is a 1 in 20 chance that we are wrong, and so on. Statistics will allow us to analyse these sorts of problem and tell us whether we can be confident enough to make our claim.

You should always avoid making claims unless you can be very confident you are right. Don't claim your batteries are better if there is a reasonable chance they are not!

The *null hypothesis* is basically to say 'Nothing has changed' – so, in our example, the new batteries are no better than the old ones. This is usually denoted by H_0. The *alternative hypothesis* is the claim we are trying to make – in this case 'The batteries do last longer than the old ones' – and is usually denoted by H_1.

If we gain the evidence that the alternative hypothesis is likely to be true, then we *reject* the null hypothesis and *accept* the alternative hypothesis ('Yes, the new batteries are better'). If we don't have enough evidence, then we *do not reject* the null hypothesis ('Not enough evidence to say the new batteries are any better').

This should become clearer when we do an example in the next chapter.

● Errors

When performing hypothesis testing, there are two ways in which you can reach the wrong conclusion.

- A *Type I* error occurs when you reject the null hypothesis in favour of the alternative hypothesis, but actually the null hypothesis is true. In our example, this means that we claim the batteries are better when actually they are not.
- A *Type II* error occurs when you do not reject the null hypothesis, but actually the alternative hypothesis was true. In our example, this means that we do not make the claim, even though it is actually true.

A good parallel to this is the British legal principle of 'innocent until proved guilty'. The onus is on the prosecution to prove, beyond all reasonable doubt, that the person is guilty – otherwise they remain innocent. Here the null hypothesis is 'innocent' and the alternative hypothesis is 'guilty'. A Type I error would be finding the person guilty when they are actually innocent, and a Type II error would be finding them innocent when actually they are guilty. Under the principle, a Type I error (you are prosecuting an innocent person) is probably considered as more serious than a Type II error (a criminal goes free), and so we should ensure that we never convict unless there is no reasonable doubt as to the guilt.

Type I errors are often known as *false positives*: you are falsely making a claim (a 'positive' statement).

> ✔ In a chapter like this, the main aim is to relate the concepts to real-life things you are aware of, to develop your understanding. Statistics is not all just learning formulae and doing maths; it's important that you understand the basic ideas. When reading material, make sure you understand what the author is trying to tell you. Are they giving you facts to learn, or are they trying to make you understand a basic concept? Both of these are needed: you need both knowledge and understanding. Always be sure what the author is trying to tell you!

● Confidence intervals

A *confidence interval* for a certain percentage is a range of values between which we expect that percentage of all observations to lie. This is easier seen by example. If you take an $N(0, 1)$ normal distribution (remember that all normal distributions can be reduced to this), then 95% of the values lie between roughly -1.96 and 1.96. How do we know this? You can calculate it from the $N(0, 1)$ table. The probability of the value lying between 0 and 1.96 is 0.4750, and similarly the probability of it lying between -1.96 and 0 is 0.4750. These probabilities add up to 0.9500, or 95%.

Graphically, this looks like the interval shown in Figure 20.1. This demonstrates a 95% confidence interval: we can be 95% confident that a value will lie in this range.

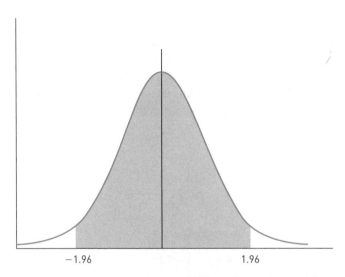

−1.96	1.96

Figure 20.1 **Example of a graph showing a 95% confidence interval for a two-tailed test**

Note that this diagram has two 'tails' (one at each end) outside the confidence interval.

A common use of confidence intervals is in testing. We might create a test for an alternative hypothesis that comes down to an $N(0, 1)$ distribution, and create a test value. We reject the null hypothesis and accept the alternative hypothesis only if the value lies in one of the tails (outside the confidence interval), giving us genuine confidence that the result really is unlikely to be from the null hypothesis, and likely (95% likely) to be from the alternative hypothesis.

A test that uses a confidence interval like the one above is called *two-tailed* for obvious reasons. There is one tail at each end, so the value is 'outside the range': either it's too big or too small.

In our example with the batteries, we claimed that the new batteries were actually better, so we should accept this claim only if they are genuinely better – not just different (either better or worse), but actually better. Therefore we are testing, for our alternative hypothesis, whether the value lies at the top end of the distribution. The only reason why we shall reject the null hypothesis and accept the alternative hypothesis is if there is genuine evidence that the average lies in the upper tail.

The diagram for a test like this looks like Figure 20.2, where there is only one tail: so a test like this is called (again for obvious reasons) *one-tailed*. The value of 1.6449 is known as a *critical value*, and can be found from Table B in the Appendix, which contains a list of the critical values and confidence intervals for a range of commonly used one-tailed and two-tailed situations.

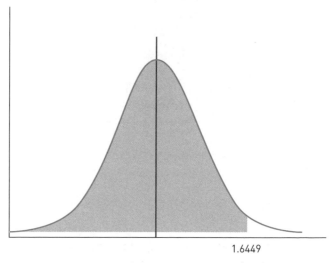

1.6449

Figure 20.2 Example of a graph for a one-tailed test showing a critical value of 1.6449

> **Remember** – a one-tailed test will be used when we want to show that one thing is greater than another thing – or, similarly, that one thing is less than another thing. A two-tailed test will be used when we simply want to show that two things are different. They can be either larger or smaller; we just want to know whether they are different.

Summary

In this chapter I've tried to get across the basic concepts of hypothesis testing to you; this will be one of the main uses of statistics you will encounter. It should become a lot clearer when we start doing examples of tests, and you'll see these concepts being used. For now, I just want you to have a basic understanding of the ideas.

Exercises

1 The marks obtained in a class test by the students were 2, 9, 10, 8, 3, 1, 0, 2, 4, 1. What was the mean mark? Take a sample of any three students you like. What is their average mark? Is this close to the actual mean? Repeat a couple of times. How representative or unrepresentative can the sample of three students be?

2 Remember that the 95% confidence interval (with two tails) was calculated by looking up the value in the $N(0, 1)$ table that had probability approximately 0.475. This value was 1.96, and so the interval -1.96 to 1.96 worked because the total probability of being in this interval is $0.475 + 0.475 = 0.95 = 95\%$. Can you work out roughly between what values the 80%, 90% and 99% confidence intervals would lie? (*If you are comfortable with it, you might like to use linear interpolation for an even more exact answer.*)

21 | z-tests

The idea of z-tests for hypothesis testing

Often we should like to test something that we believe follows a
normal distribution. If we know the variance and standard deviation
of this normal distribution, then a z-test is a powerful technique for
hypothesis testing.

Key topics

● z-test

Key terms
z-test test statistic rejection confidence

Recall that when we discussed normal distributions, if x is normally
distributed with mean μ and standard deviation σ, then the variable
$z = \frac{x - \mu}{\sigma}$ follows the normal distribution $N(0, 1)$. The use of the
letter z is so common with normal distributions that tests that
involve the normal distribution are given the general name z-*tests*.

● The z-test statistic

z-tests are generally used to test properties of means, such as in the
example we gave in the previous chapter. Does a sample mean of 7.1
(compared with a mean of 7 for the competitor) convincingly prove
that the batteries are genuinely better?

What I am going to do is present the basic theory, and then a
formula we shall use. I shall then give examples of using this
formula.

Given a population, there are lots of different samples you could take of a certain size n, and all of these will probably have different means, which we hope are close to the true mean of the population. It is possible to show that if the population is normally distributed, with a distribution of $N(\mu, \sigma^2)$, then the means of all samples of size n are also normally distributed, with distribution $N(\mu, \frac{\sigma^2}{n})$, so the mean is μ, and the standard deviation is $\frac{\sigma}{\sqrt{n}}$ (the square root of the variance $\frac{\sigma^2}{n}$), which I shall write on one line as $\sigma/\sqrt{n}$ to make the next formula a little easier to read.

Essentially, the idea of our check will be to see whether our sample mean is what we should reasonably expect, if it came from our distribution.

To do this, remember that to transform a normal variable into the $N(0, 1)$ distribution, we subtract the mean and divide by the variance. We are going to use this basic idea to create our test variable, which will be more formally referred to as the *test statistic*. The following is the most important thing you need to know:

If a population has mean μ and standard deviation σ, and we have a sample of size n and mean $\overline{x}$, then the *test statistic* z is given by

$$z = \frac{\overline{x} - \mu}{\sigma/\sqrt{n}}.$$

This is really all you need to know, although the basic mathematical idea behind it is useful. Let's see how it works in practice.

● Example: A one-tailed z-test

Let us step carefully through how to perform a z-test. I'll describe the steps in detail, although in practice you would probably present your argument rather more concisely – which I'll do for the second example, but for now it's more beneficial to talk through the full details. We'll use something similar to the batteries example from before.

Scenario

A company has developed a new range of batteries. They test a sample of 50 of these batteries, and find that they have a mean life of 7.1 hours. Their older range of batteries was known to follow a normal distribution, with a mean of 7 hours and a standard deviation of 0.35. The company claims their new batteries are longer-lasting than the old ones. Is this claim justified?

Step 1: Write down the null hypothesis and the alternative hypothesis

The null hypothesis is that nothing has changed: that is, the batteries are just the same as before. This means that the mean is still 7. The alternative hypothesis is that they have improved, and so the mean is now more than 7. Using μ for the mean as usual, we can write this formally as:

$$H_0: \mu = 7$$
$$H_1: \mu > 7$$

Remember that we are assuming that H_0 is true, and we shall need considerable evidence to change our mind and reject the null hypothesis H_0 in favour of the alternative hypothesis H_1.

Step 2: Create the z-test statistic

We have $\mu = 7$ from the null hypothesis (we are assuming H_0 is true, so $\mu = 7$). Also we have $\sigma = 0.35$ from the data given in the scenario. The mean of our sample was 7.1, and so we have $\bar{x} = 7.1$, and $n = 50$ since we have a sample size of 50.

Then our test statistic is $z = \dfrac{\bar{x} - \mu}{\sigma/\sqrt{n}} = \dfrac{7.1 - 7}{0.35/\sqrt{50}} = 2.0203$

(to 4 decimal places) using a calculator.

Step 3: Decide how confident you need to be in your answer

Soon we shall test this 'test statistic' value, but first we have to decide how confident we need to be in our answer. Common values are 95% and 99% confident. We'll do this first with 95% confidence. This is also referred to as 'testing at the 5% level', since there is only a 5% chance that we are wrong.

As discussed in Chapter 20, this means that we need the *critical value* to be the value whereby any value of z above this will mean that we reject the null hypothesis. In this case we have a one-tailed 95% test, and so from Table B this value is around 1.6449.

Step 4: Make a decision as to whether to reject the null hypothesis.

Our test statistic is 2.0203, which is more than 1.6449. It's therefore very likely (we can be 95% sure) that this sample came from a distribution with a higher mean than 7, and so we are confident enough at the 5% level to reject the null hypothesis and accept the alternative hypothesis. So we are confident to make the claim that these batteries are better.

Step 5: Make a formal conclusion

You should always state a formal conclusion. In this case, because of our confidence figure of 95%, we can conclude that there is good evidence that these batteries are better than the old ones.

A similar example

Next we shall do almost exactly the same thing, but this time test with 99% confidence, or at the 1% level. I shall simply summarise the steps and write the mathematics down. At first you should write down as much you need, but in time you can get used to just writing down the important things, as here. The only difference in the calculation is the value for the 99% confidence limit, which is read from Table B in the Statistical Tables section.

Hypotheses

The null hypothesis H_0: $\mu = 7$

The alternative hypothesis H_1: $\mu > 7$

z-test statistic

$$z = \frac{\bar{x} - \mu}{\sigma/\sqrt{n}} = \frac{7.1 - 7}{0.35/\sqrt{50}} = 2.0203 \text{ (to 4 decimal places)}$$

Critical value

A one-tailed 1%-level test, so the critical value is 2.3263.

Decision

Since the test statistic 2.0203 is less than the critical value, we do not reject the null hypothesis.

Conclusion

There is not enough evidence to confirm the claim at the 1% level, so we cannot make the claim with very strong confidence.

So, while we can be confident at the 5% level, we cannot be confident at the 1% level. We might say that we have 'some evidence' for the claim, but we do not have 'strong evidence'. These are quite common terms for the 5% and 1% levels.

Most of the time, when we have questions like this we shall use one-tailed tests, since we usually want to confirm whether one thing is 'better' or 'worse' than another. If you do come across the need for a two-tailed test, which is where you might just want to say that a sample 'does not come from' a population, regardless of whether the mean is too high or too small, then you will need to do the same thing, but just use the two-tailed figures from Table B.

● Example: Comparing means with a two-tailed test

A common use of a z-test is to compare two populations and see whether there is any significant difference between their means. This can be used, for example, in medical trials, where you would like to compare the effectiveness of two drugs, by using two sample groups.

Effectively, what we are doing here is testing the difference of the two means. If the difference is small, then there is no evidence to suggest that one group is significantly different from the other, whereas if the difference is large, there may be a statistically significant difference between the groups.

The main fact (we won't prove this) is that if we have a sample of size n_1 from a normal distribution with mean μ_1 and standard deviation σ_1, and another sample of size n_2 from a normal distribution with mean μ_2 and standard deviation σ_2, then the difference between the means is a normal distribution with mean $\mu_1 - \mu_2$ and standard deviation $\sqrt{\dfrac{\sigma_1^2}{n_1} + \dfrac{\sigma_2^2}{n_2}}$. This makes our test statistic formula rather complicated. This isn't easy to learn, and I hope will be given for you, but if you *do* have to learn it, note that it's similar to the formula we saw before, but we have two of everything now.

If $\bar{x}_1$ is the mean of a sample of size n_1 from a normal distribution with mean μ_1 and standard deviation σ_1, and $\bar{x}_2$ is the mean of a sample of size n_2 from a normal distribution with mean μ_2 and standard deviation σ_2, then the z-test statistic is

$$z = \frac{(\bar{x}_1 - \bar{x}_2) - (\mu_1 - \mu_2)}{\sqrt{\dfrac{\sigma_1^{\,2}}{n_1} + \dfrac{\sigma_2^{\,2}}{n_2}}}$$

Make sure you know what you have to learn, and what will be given to you. If you are a student revising for an exam, there is no point in spending hours learning complex formulae like this if you know that in your exam it will be provided!

Now let's see this in action. We shall take the example of two classes at different universities, studying the same course, and investigate whether their results are statistically significantly different. We shall use a two-tailed test, as we aren't checking whether one university is 'better' than another, just whether the two universities are significantly different, in either way (better or worse). You could also ask the question whether one university is 'better' than the other, so it becomes a one-tailed test. You could do this as an exercise.

Scenario
The marks for a sample group of 30 students at the course at University A have a mean of 58 and a standard deviation of 4.9. The marks for a sample group of 50 students on the course at University B have a mean of 55 and a standard deviation of 5.1. Is there evidence of a significant difference between the standards of the two universities? Use a 5% level.

Hypotheses
The null hypothesis is that the universities are of the same standard, so the two means should be equal (and any difference is down to chance). The alternative hypothesis is that they have different standards, so the two means are significantly different (not just due to natural chance, but a fundamental difference). Formally, this is

Null hypothesis H_0: $\mu_1 = \mu_2$
Alternative hypothesis H_1: $\mu_1 \neq \mu_2$

z-test statistic

With so many values, it's important that we are careful to get everything right, so first make a systematic list of what each of the variables are. We have:

Sample means: $\bar{x}_1 = 58, \bar{x}_2 = 55$

Standard deviations: $\sigma_1 = 4.9, \sigma_2 = 5.1$

Size of the samples: $n_1 = 30, n_2 = 50$

Also, although we don't of course know what the actual means μ_1 and μ_2 are, the null hypothesis says that $\mu_1 = \mu_2$ (we are assuming them to be the same), so we have $\mu_1 - \mu_2 = 0$.

Hence the z-test statistic is

$$z = \frac{(\bar{x}_1 - \bar{x}_2) - (\mu_1 - \mu_2)}{\sqrt{\dfrac{\sigma_1^2}{n_1} + \dfrac{\sigma_2^2}{n_2}}} = \frac{(58 - 55) - 0}{\sqrt{\dfrac{4.9^2}{30} + \dfrac{5.1^2}{50}}}$$

which works out to be 2.6106 to 4 decimal places.

Confidence intervals

We are using a 5% two-tailed test, so our confidence interval is from -1.96 to 1.96 (looked up from Table B).

Decision

We have to have either $z < -1.96$ or $z > 1.96$ to reject the null hypothesis. Since $z = 2.6106$, we do have $z > 1.96$, and so we reject the null hypothesis and accept the alternative hypothesis.

Conclusion

There is evidence at the 5% level that the universities do have different standards.

As before, let's do this again at the 1% level, just summarising the steps. Remember that the only difference is the confidence interval.

So in this example we have strong evidence that the universities do differ in standards, and it's not just down to chance.

 ## Summary

z-tests are a powerful technique, used to determine whether we have enough evidence to make a claim. Remember that we should make the claim, and so reject the standard situation (the null hypothesis), only if we can make the claim with confidence. Unfortunately, z-tests do rely on you knowing the standard deviation, which is often not known. We shall look at this in the next chapter, when we consider a closely related type of test called a t-test.

 ## Exercises

Perform a z-test on the following claims at both the 5% and 1% level. You will have to decide whether you need a one-tailed or a two-tailed test, and so use the appropriate figures from Table B.

1 The life expectancy of a particular type of light-bulb is known to have mean 100 hours with standard deviation 5 hours. A newly developed bulb is tested, and from 40 tests the mean is 101.5 hours. The manufacturers claim the new bulb lasts longer. Is this claim valid?

2 The time taken for a particular seed to germinate is known to be 13.5 days, with standard deviation 0.5. Fertiliser manufacturers test a new fertiliser on 50 seeds, and the average germination time is 13.4 days. They claim that their fertiliser is effective in reducing the germination time. Is this claim valid?

3 Two new beauty products have been trialled. The first product was tested on a sample of 50 people, and their average quality rating was 6.9 out of 10, with standard deviation 0.7. A second product was tested on a sample of 60 people, and the average quality rating was 6.6, with standard deviation 1.1. Is there any evidence to suggest that the two products are of different· quality?

4 A sample of 40 students who have read a particular textbook gives a mean mark of 6.1 and standard deviation 1.3. Another sample of 40 students who have not read the book gives a mean mark of 5.2 and a standard deviation of 1.1. The publishers of the textbook claim that these results show that reading the book will give better marks. Is this a valid claim to make?

The use of *t*-tests when the standard deviation of the population is unknown

Key topics
● *t*-test

Key terms
sample standard deviation *t*-test statistic paired *t*-test
unpaired *t*-test

With the *z*-test, we knew what the standard deviation of the population was. Unfortunately this isn't often the case. If I asked a random person for the average height of a UK male they might not be far off in their guess, but if I asked them the standard deviation, they might well just look blankly at me, and have no idea.

When we don't know the standard deviation, the best we can do is to estimate it from the standard deviation of the sample we have taken, and then test with that.

● Sample standard deviation

Recall that for the *sample* standard deviation we do the same thing as for the *population* standard deviation, but divide by $n - 1$ instead.

Using our sigma notation, the formula is $\sqrt{\dfrac{\sum\limits_{i=1}^{n}(x_i - \bar{x})}{n - 1}}$. Do remember this difference. The $n - 1$ is needed whenever we have only a sample (as noted before, it's beyond the scope of this book to go into detail why).

The sample standard deviation is of course only an approximation to the actual standard deviation of the population. The larger the sample, the closer it is likely to be. Once the sample becomes reasonably large (30 is a good guide), then it's likely to be close enough to be taken as the actual standard deviation, and we can use a z-test with confidence, but for smaller sample sizes we need a different approach.

> **i** Remember, in all of this, we do not know the standard deviation of the whole population, just that of the sample. This is the crucial difference between a t-test and a z-test.

● The t-test statistic and Student's t-distribution

The smaller the sample size, the less confident we can be in a test, since the more unsure we are as to whether our estimate for the standard deviation is close to the actual standard deviation. Hence, for every different sample size, we shall need different confidence limits.

The main statistic for a t-test is very similar to that for a z-test, and can be defined as follows:

> **i** The *t-test statistic* for a sample of size n, with mean x and sample standard deviation s, from a population with known mean μ (remember, we don't know the population standard deviation), is given by $t = \dfrac{\bar{x} - \mu}{s / \sqrt{n}}$.

We are going to read off our confidence limits from a table as before, but remember that there need to be different confidence limits for different values of n. The distribution for a particular value of n is known as *Student's t-distribution* with $n - 1$ *degrees of freedom*. This is a rather strange name (Student): the creator of the distribution was William Gosset, who worked for a company that would not allow their staff to publish scientific material in case they revealed trade secrets, and so he adopted the simple pseudonym 'Student'. *Degrees of freedom* is also rather a complex name. It boils down to the number of variables you have (how 'free' you are to choose things), but again this is beyond what we can discuss here. The most important thing you need to remember is:

For a sample of size n, the number of 'degrees of freedom' is $n - 1$.

● t-tables

Just as with the z-test, there are tables that can be used for various confidence limits. These are given in Table C1 and Table C2 in the Statistical Tables section, which are for one-tailed and two-tailed tests respectively. To use these tables, you need to find the number of degrees of freedom (remember, this is one less than the sample size) and then look along the row for the confidence limit you require; I've included the most common ones. Have a quick look at these tables, but we'll do a few examples now to illustrate their use.

● Example of a t-test: comparing a sample mean

One use of a t-test is comparing a sample mean with a mean already known.

Suppose, for example, that a supermarket sells bags of potatoes, all of which are labelled as 2.5 kg and are of an equal price. These bags are supplied from a local farm. Of course, the weight of bags can vary a little, but the farm charges exactly per kilogram, so a 2.6 kg bag is more expensive than a 2.4 kg bag, for example.

The average weight of a bag of potatoes supplied should be exactly 2.5 kg. It should not be more than this, or the supermarket will lose money, since they will be paying for more than 2.5 kg but selling them at the 2.5 kg rate. And it should not be less, or customers will be dissatisfied with their bags.

A new supplier approaches the supermarket and offers to supply potatoes a little cheaper per kilogram. The local farm pleads with the supermarket to keep them as their supplier, and asks them to test the supplier and only switch if they are really certain their bags are 2.5 kg mean.

The new supplier agrees, and gives a sample of 20 bags of potatoes to the supermarket for testing. The supermarket finds that the mean weight of these 20 bags is 2.55 kg with sample standard deviation 0.1 kg. Should the supermarket use this new supplier, or is there evidence from this sample that the bags of potatoes from this supplier have a mean different from 2.5 kg?

Let us first do this with a 5% confidence level. I shall go through all the details (which are very similar to before), and then we'll do a 1% test more concisely.

Variables
First write down what we know. The mean μ must be 2.5, but we don't know what the standard deviation must be, so we need a t-test. Our sample mean is $\bar{x} = 2.55$ and our sample standard deviation is $s = 0.1$. Also, we have $n = 20$, since there are 20 bags in the sample.

We shall need to do a two-tailed test, since we want to check that the mean is 2.5 kg, and should not be either bigger or smaller. So we shall use Table C2.

Hypotheses
The null hypothesis is that the mean of the new suppliers is indeed 2.5 kg, and the alternative hypothesis is that the mean is actually different. Using the usual notation, we have:

$H_0: \mu = 2.5$
$H_1: \mu \neq 2.5$

Remember that we are assuming that H_0 is true, and we shall need considerable evidence to change our mind and reject the null hypothesis H_0 in favour of the alternative hypothesis H_1.

t-test statistic
Using the variables as we defined, we have that the t-test statistic is
$$t = \frac{\bar{x} - \mu}{s\sqrt{n}} = \frac{2.55 - 2.5}{0.1/\sqrt{20}} = 2.2361 \text{ (to 4 decimal places)}.$$

Confidence interval
We shall start by doing this at the 5% limit, so we can be 95% confident in our answer.

We have $20 - 1 = 19$ degrees of freedom (remember, this is one less than the sample size), so we use Table C2 (since it is a two-tailed test) and look up the 5% limit for the {degrees of freedom $= 19$} row, which gives 2.093 from the table. This means that the confidence interval is between -2.093 and 2.093, and so we shall reject the null hypothesis if we lie outside this range.

Make a conclusion

Our test statistic is 2.2361, which is outside the range, and so we have enough evidence at the 95% level. So we reject the null hypothesis and accept the alternative hypothesis, that the mean is different, and stay with our current local farm.

We can do this test at the 1% level as well, which means that we need to be 99% sure of our decision that the mean is different. This calculation is virtually identical, and is summarised below.

Variables

As above, $\mu = 2.5$, $\bar{x} = 2.55$, $s = 0.1$ and $n = 20$, and we shall do a two-tailed test.

Hypotheses

As before, we have H_0: $\mu = 2.5$ and H_1: $\mu \neq 2.5$.

***t*-test statistic**

Using the variables as we defined, we have the *t*-test statistic is

$$t = \frac{\bar{x} - \mu}{s\sqrt{n}} = \frac{2.55 - 2.5}{0.1/\sqrt{20}} = 2.2361 \text{ (to 4 decimal places)}.$$

Confidence interval

There are $20 - 1 = 19$ degrees of freedom. Looking up the 1% limit in Table C2, the value is 2.861, and so the confidence interval is between -2.861 and 2.861.

Conclusion

Our test statistic is 2.2361, which is inside the range, and so we do not have enough evidence at the 99% level that the means are different, so we stay with the null hypothesis that the means are the same, and therefore can switch suppliers.

Overall, there is some evidence (5% level) that the means are different, but there is not very strong evidence (1% level), and so we might conclude that there is not an overwhelming case for a difference in mean. So we might decide to go with the new supplier; although there's some evidence that the average weight might be different from 2.5 kg, it's not conclusive, so we may well just go with it – although this is a decision that someone has to make!

> ✓ Don't forget that the statistical test is just giving you information. In this case we can be reasonably sure, but not very sure, that the mean is different – so should we switch to the cheaper supplier? This is now a decision for a human to make, but at least they have the figures available to help them make the decision.

● Paired *t*-test

A common use of a *t*-test is in comparing two methods on a sample. Suppose, for example, we take 10 patients on a hospital ward who suffer from an illness that causes short-term symptoms, for which they take medication. The time taken for their symptoms to be relieved, after their medication is taken, is given in the table below.

Patient	A	B	C	D	E	F	G	H	I	J
Time	9	4	7	10	12	3	2	18	5	1

A new drug is being trialled that the makers claim shorten the time for the symptoms to be relieved. The 10 patients try this drug, and their recovery times are:

Patient	A	B	C	D	E	F	G	H	I	J
Time	7	2	5	8	8	2	3	12	2	1

Is there evidence to suggest that the drug really does decrease recovery time?

This is an example of *paired* data. For each patient, there is a 'pair' of data: the time for the old (currently used) drug and the time for the new (being trialled) drug.

We can perform a *t*-test from this sample to see whether the new drug really is any better. The idea is to look at the differences between the old treatment and the new treatment for each patient. If the average of these is more than 0, then this means that the average difference is more than zero, which implies that the new drug is better: on average the treatment time is reduced as, on average, the difference between the old drug and the new drug is positive.

We can create a table of differences:

Patient	A	B	C	D	E	F	G	H	I	J
Old	9	4	7	10	12	3	2	18	5	1
New	7	2	5	8	8	2	3	12	2	1
Difference	2	2	2	2	4	1	-1	6	3	0

I shall leave it for you as an exercise to work out that the mean of the differences is 2.1 and the sample standard deviation (don't forget that you need to divide by $n - 1 = 10 - 1 = 9$) is approximately 1.9692.

Now we can perform a t-test. I shall do it at the 1% level (and so a 99% confidence interval), but you should try the calculation for yourself at other levels. Remember that we want to test whether the new mean is greater than 0.

Variables

We have $\mu = 0$ (this is what we are using as the test for what the mean is), $x = 2.1$, $s = 1.9692$ (from what we just calculated), and $n = 10$ (sample size is 10, there are 10 patients), and we shall do a one-tailed test (since we are testing whether the mean is greater than 0).

Hypotheses

The null hypothesis is that the new drug is no better, so the average difference is zero, while the alternative hypothesis says that it is better (the average difference is more than zero), and so we have have $H_0: \mu = 0$ and $H_1: \mu > 0$.

t-test statistic

Using the variables as we defined, we have that the t-test statistic is

$$t = \frac{\bar{x} - \mu}{s\sqrt{n}} = \frac{2.1 - 0}{1.9692 / \sqrt{10}} = 3.372 \text{ (to 3 decimal places)}.$$

Critical value

There are $10 - 1 = 9$ degrees of freedom, and so looking up the 1% limit in Table C1 (one-tailed tests), the critical value is 2.821.

Conclusion

Our test statistic is 3.372, which is above the confidence limit, and so we have evidence at the 1% level that the mean difference is greater than zero. So we reject the null hypothesis, and accept the alternative hypothesis (that the drug is better) with very strong confidence.

● Unpaired t-test

In the example above we tried both of the drugs on the same patients, and so the results fell into 'pairs': two results for each patient, one with the old drug and one with the new drug. However, it's also very common to have data from two independent samples – so not trying both drugs on all people in the sample, but trying one drug on some of the people, and the other drug on the rest of them.

You can do a t-test here as well. The hardest part is in creating the t-test statistic, as the formulae can get quite complicated, although once you have evaluated the statistic, you just test it as usual.

As the values aren't paired there's no way we can take differences, but what we *can* do is check to see whether the means of the two samples differ. For simplicity I'll first do an example where the sample sizes are equal, before generalising to cases where we have different sample sizes.

Unpaired t-test with equal sample sizes

A trial is taking place in which eight people are given only water, whereas another group of eight people are given a new energy drink. They then have to take part in an endurance test. The results of the trial are given in the following table, which records the mean and standard deviation of the performance of the two groups, in terms of the length of time they were able to perform the endurance task for. Is there significant evidence that the people who have taken the energy drink perform better?

	Mean	Standard deviation
Water	12.2	2.4
Energy drink	13.1	3.1

We can do a t-test here again, but we need to work out the overall standard deviation of the sample. This is called a *pooled estimate* of the standard deviation, and is obtained by averaging the variances amongst the whole group. Remembering that the variance is the square of the standard deviation, it makes sense to define the pooled estimate of the standard deviation to be $s_p = \sqrt{\dfrac{s_1^2 + s_2^2}{2}}$ where s_1 and s_2 are the two standard deviations (the p in s_p is short for 'pooled').

Hence in our example we have $s_{\text{p}} = \sqrt{\dfrac{2.4^2 + 3.1^2}{2}} = 2.7722$ (to 4 decimal places).

Let us now perform a t-test on this data, to see whether there is sufficient evidence from these samples that the energy drink influences performance.

Quite similarly to the z-test, the t-test statistic will be given by $t = \dfrac{(\bar{x}_1 - \bar{x}_2) - (\mu_1 - \mu_2)}{s_{\text{p}}/\sqrt{n/2}}$, where $\bar{x}_1$ and $\bar{x}_2$ are the sample means, μ_1 and μ_2 are the population means (which we don't know, but as you'll see in a moment, we are going to test that they are equal, and so we have $\mu_1 - \mu_2 = 0$), and n is the sample size of each group (8 in our case).

Also, we need the number of degrees of freedom. Remembering that this is one less than the sample size, we have 7 degrees of freedom in the first group and 7 degrees of freedom in the second group, and so overall there are 14 degrees of freedom. In general, for sample sizes of size n, the number of degrees of freedom will be $2n - 2$ (this comes from $(n - 1) + (n - 1) = 2n - 2$).

Let's do the calculation, which should make things clearer:

Variables

$\bar{x}_1 = 12.2$ and $\bar{x}_2 = 13.1$, we worked out $s_{\text{p}} = 2.7722$ and $n = 8$ (sample size of each group is 8), and we shall do a one-tailed test, since we are testing whether the mean of the people using the energy drink is greater than that of the people using only water. Let us do a test at the 5% limit (so with 95% confidence).

Hypotheses

The null hypothesis is that the energy drink makes no difference, and so the population means are the same, while the alternative hypothesis says that the mean of the first population is less than that of the second, and so we have the following.

$$H_0: \mu_1 - \mu_2 = 0$$

$$H_1: \mu_1 - \mu_2 < 0$$

t-test statistic

Using the variables as we defined (remember that we are assuming the null hypothesis, so $\mu_1 - \mu_2 = 0$), we have that the t-test statistic is

$$t = \frac{(\bar{x}_1 - \bar{x}_2) - (\mu_1 - \mu_2)}{s_{\text{p}}/\sqrt{n/2}} = \frac{(12.2 - 13.1) - 0}{2.7722/\sqrt{(8/2)}} = -0.6493 \text{ (to 3 decimal places)}$$

Unpaired *t*-test with unequal sample sizes

You can perform an unpaired *t*-test when the sample sizes are different; it's almost exactly the same as above, but the formula is more complicated.

I won't go through an example, as it's not really any different from what we did before, but I'll give the formulae for your reference.

If you have sample sizes of n_1 and n_2, then the pooled standard deviation estimate is

$$s_p = \sqrt{\frac{(n_1 - 1)s_1^2 + (n_2 - 1)s_2^2}{(n_1 - 1) + (n_2 - 1)}}$$

where s_1 and s_2 are the two standard deviations, as before.

The *t*-test statistic is then

$$t = \frac{(\bar{x}_1 - \bar{x}_2) - (\mu_1 - \mu_2)}{s_p \sqrt{\frac{1}{n_1} + \frac{1}{n_2}}}$$

I deliberately didn't do an example here, as it's just what we've done before, but with a more complex formula, and I didn't want this book to look too daunting to someone glancing casually at it. If you do need to do questions like this, you should find or create some examples of your own and work through the calculations. You have all the knowledge you need to solve this sort of problem; you just need to put numbers in the formula!

The formulae involved can get even more complicated if the two samples come from populations with different variances, but we won't discuss this here.

Summary

In most cases we don't know the standard deviation of the population, and so we have to estimate from the sample: this is the main difference between a t-test and a z-test. The actual testing procedure is very similar, and is probably the most important test that you will encounter.

Exercises

You should answer these questions at both the 5% and 1% levels. You will need to decide whether you need a one-tailed or two-tailed test.

1 The average heart rate during a stressful event is known to have mean 100. In a trial, 12 volunteers test a pill that claims to lower the heart rate. The mean heart rate in the sample is 95, with standard deviation 6. Perform a t-test on the claim that the heart rate is lower, based upon this sample.

2 Eight people take part in a reaction test, and their speed of response is measured before and after eating a cereal bar. The figures are below:

Person	A	B	C	D	E	F	G	H
Before	12	10	8	4	3	22	6	9
After	9	10	4	5	2	24	5	7

By calculating the differences, and then working out the mean and standard deviation of the differences, perform a paired t-test to see whether there is any evidence that eating the cereal bar has any effect on reaction time.

3 Ten plants were left to grow naturally, and 10 plants were treated with a new fertiliser. The mean height of the first sample (which grew naturally) was 7.8 cm, with standard deviation 1.2 cm, and the mean height of the second sample (treated with the fertiliser) was 8.7 cm, with standard deviation 0.9 cm. Use an unpaired t-test to determine whether the fertiliser has an effect on height.

23 | χ^2-tests

The use of χ^2 (chi-squared) tests for observed and expected values

Often you have an expectation as to how data will fall in various classes. If you get a set of data, does it match this expectation? We can use a χ^2-test to see whether the data fits our expectations, or is significantly different.

Key topics
● χ^2-test

Key terms
observed data expected data classes χ^2-test contingency tables
degrees of freedom Yates' correction

The tests we talked about in the last two chapters (the z-test and t-test) were concerned with comparing means. But comparing means isn't all you want to do with statistical data. In this chapter we shall look at what happens when data falls into various classes, and compare it with what we expected to happen.

● Observed and expected data in classes

Data often falls into classes. An example would be the classification of exams, where students are given an A, B, C, etc. grade.

Often there is an expectation as to how the data might fall into the classes. For example you might say that, ideally, we should expect 10% of students to get an A grade, 20% to get a B grade, 40% to get a C grade, 20% to get a D grade, and the remaining 10% an E grade.

If we then get the actual results, we want to be able to compare the actual frequencies (how many were in each category) with this expectation, to know whether they differ significantly from what was expected. If the results are significantly different, then perhaps we should investigate: was the exam paper too easy, or too hard, or was there a change in teaching standard? Performing a test on the data will tell us whether the data matches the expected pattern reasonably well, or whether there is cause for concern.

As always, we can test at different levels (such as 5% or 1%), depending on how certain we need to be of our conclusion.

● The χ^2-test statistic

Suppose the *observed frequencies* (the actual data we obtained) are denoted by O_i and the *expected frequencies* (what we expected to happen) are denoted by E_i. The idea is to give a measure of the difference between the observed frequencies and the expected frequencies. The common measure is what is known as the χ^2-test statistic. The symbol χ is a Greek letter, which can be written in English as 'chi': this is pronounced roughly as 'ky' (to rhyme with 'sky').

> Be aware of how to search for help on the Internet. You almost certainly don't have a button for χ^2 on your keyboard, but if you know that it is referred to as 'chi-squared' using normal English letters, then you can easily search for it using online search engines. Make sure you know how to search for the things you need when they involve strange symbols. Note that some authors prefer 'chi-square' rather than 'chi-squared', so do be aware when you're searching that people sometimes use slightly different terms.

The χ^2-test statistic is given by:

$$\chi^2 = \sum_{i=1}^{k} \frac{(O_i - E_i)^2}{E_i}, \text{ where } k \text{ is the number of classes}$$

What this says is: for each value, take the difference, square it, and divide by the expected value, and then sum all of these together.

Note that this statistic will never be a negative number, as we are counting how many things fall into each class (which can't be negative), and the differences are squared, so they can't be negative either. Hence the χ^2-test statistic will always be positive (or zero). We shall test as usual: we shall give the null and alternative hypothesis, calculate this χ^2-test statistic, and find whether it lies outside a range given by tables to see whether we have enough evidence to accept the alternative hypothesis. Since it will never be negative, we can only ever use a one-tailed test.

We are also going to use the phrase *degrees of freedom* again, as we did with the t-test. I shall explain the relevant degrees of freedom as we go through the examples. For a simple test with five categories (A, B, C, D, E grades, say), the number of degrees of freedom is one less than the number of categories: so $5 - 1 = 4$.

If there was no difference between the observed frequencies and the expected frequencies, then the χ^2-test statistic would be 0, since there are no differences, and so every term in the sum is 0. Hence, almost always, you will find that the null hypothesis is that $\chi^2 = 0$, and the alternative hypothesis is $\chi^2 > 0$. Remember, we shall need significant evidence to reject the null hypothesis and conclude that there is a significant difference from the expected frequencies, and something might be wrong that we should investigate.

As always, an example will make things clearer.

● χ^2-test: Example with one criterion

As a first example, we shall consider the exam marks we discussed before. Here we are classifying the students according to one criterion only – their grade. We'll do an example later where there is more than one criterion.

Remember, we are expecting 10% of students to get an A grade, 20% to get a B grade, 40% to get a C grade, 20% to get a D grade, and the remaining 10% to get an E grade. Significant differences from this would cause us to have some concerns, and perhaps investigate.

From 200 students, the actual figures were as follows:

A: 32   B: 48   C: 71   D: 30   E: 19

Are these consistent with what was expected, or is there a significant difference?

Before you start a question, have a quick look at the data, and form an initial idea; this will help you see whether your final answer is sensible. In this case, we do seem to have a few more As and Bs than we would expect, so perhaps the exam was rather easy. We'll have to see whether the analysis bears this out, but it's nice to have an overall idea just by glancing at something; the statistics are then meant to back up these informal ideas.

We shall create a table that will allow us to do a χ^2-test. First of all, we need to work out what we would expect. There are 200 students, and we expect 10% of these to get an A grade, so we expect 20 to get an A. Similarly, we expect 40 to get a B (20% of the 200), and so on. This gives us the following table, in which we record the actual observed values (denoted by O_i), and the expected values (denoted by E_i).

Observed (O_i)	32	48	71	30	19
Expected (E_i)	20	40	80	40	20

Now we can add rows to the table to help us with our calculation. We first work out the differences $(O_i - E_i)$, then work out these differences squared $(O_i - E_i)^2$, and finally divide each one by E_i. For example, in the first column, we work out the difference $(32 - 20) = 12$, then square this to get $12^2 = 144$, and finally divide by E_i (which is 20), to get $144/20 = 7.2$. Doing the same thing for all columns, we get the following table; check these calculations for yourself.

Observed (O_i)	32	48	71	30	19
Expected (E_i)	20	40	80	40	20
$(O_i - E_i)$	12	8	-9	-10	-1
$(O_i - E_i)^2$	144	64	81	100	1
$(O_i - E_i)^2 / E_i$	7.2	1.6	1.0125	2.5	0.05

Now, to work out the χ^2-test statistic, remember that we need to add up all the $(O_i - E_i)^2 / E_i$, and so we just add up the final row of the table to get $7.2 + 1.6 + 1.0125 + 2.5 + 0.05 = 12.3625$. This is our χ^2-test statistic.

If the observed values met the expected values exactly, we'd expect this to be zero, and it's not; but is this significant? We shall do the χ^2-test, which we present in the usual way. Remember that we are testing to see whether this value is significantly different from the 0 that we would expect. There are bound to be natural fluctuations, and the figures are unlikely to match exactly, but is the difference significant?

Hypotheses

Our null hypothesis will be that there is no significant difference, so that the statistic is 0, and any difference is just random fluctuation. The alternative hypothesis is that the statistic is greater than 0. Remember: we need significant evidence to accept the alternative hypothesis and conclude that the results are significantly different from what would be expected. So we have:

Null hypothesis: $\chi^2 = 0$
Alternative hypothesis: $\chi^2 > 0$

Test statistic

As above, we calculated the χ^2-test statistic to be 12.3625.

The number of degrees of freedom is given by one less than the number of categories, so, since we have five categories (A, B, C, D, E) the number of degrees of freedom is 4.

Critical value

Remember that χ^2 cannot be negative, so we have to do a one-tailed test. We shall obtain a critical value from tables: if our test statistic is higher than the critical value at a particular level, we reject the null hypothesis at that level.

We can look up the values corresponding to 4 degrees of freedom in Table D in the Statistical Tables section.
At the 5% level, the critical value is 9.488.
At the 1% level, the critical value is 13.277.

Conclusion

At the 5% level, since the test statistic 12.3625 is greater than the critical value 9.488, then we have enough evidence at this level to reject the null hypothesis, and accept the alternative hypothesis that there is a significant difference from the expected values, and we should investigate.

At the 1% level, since the test statistic 12.3625 is less than the critical value 13.277, then we do not have enough evidence at this level to reject the null hypothesis.

Overall, there is some evidence (but not very strong evidence) that the results are significantly different from what we might expect, and so we should probably investigate here.

Note that the result of the χ^2-test tells us nothing about where the discrepancy arises (are the marks too high, or too low, is one grade completely unexpected, etc?) but at least we have flagged that something is not as we expected, and we should have a detailed look.

Contingency tables

In the previous example we classified the data according to one criterion, which was what grade the students got, from A to E. Often, though, things are classified with more than one criterion. For example, a batch of peppers may be classified both according to their colour (say red, green, yellow, orange) and their size (perhaps small, medium, large): there are large red peppers, medium orange peppers, and so on.

Where there are two criteria, you can present the data in a *contingency table*. This is nothing more than a straightforward table of the data, split into the two criteria. It is usual to include the sum of each row and column as part of the table, for reasons that we will see shortly. The following is an example of a contingency table, which shows, for a batch of 500 peppers, the number of peppers falling into each category.

	Small	Medium	Large	Sum
Red	30	70	60	160
Green	50	100	80	230
Orange	15	30	15	60
Yellow	15	20	15	50
Sum	110	220	170	500

We can ask the question as to whether there is any connection between colour and size. For example, are green peppers more likely to be larger, or is the distribution just standard, and there is no particular connection between colour and size?

This isn't straightforward to answer. There are more medium-sized peppers than small or large ones, and there are more green peppers than any other, but that doesn't tell us anything about a connection between the two things (colour and size).

To see whether there is any connection (correlation) between the two things, we need to do a statistical test; a χ^2-test can be used here too.

The number of degrees of freedom of a χ^2-test with two criteria like this is given by $(m-1)(n-1)$, where m is the number of rows and n is the number of columns. So here, for example, we have 4 rows and 3 columns, and so the number of degrees of freedom is $(4-1) \times (3-1) = 3 \times 2 = 6$.

● χ^2-test: Example with two criteria

There are 160 red peppers, split into the ratio 30:70:60, there are 230 green peppers, split into the ratio 50:100:80, and so on. Is there any evidence that different-coloured peppers have significantly different ratios of small:medium:large? That is, is there evidence that the colour and size are related somehow, so that different colours have significantly different size ratios?

What we shall do is compare what we expect to happen, given the sample we have. The sums are critical here: we know there are 160 red peppers, but do they fall in the ratio we would expect, and so on?

To work out the expected value for a particular entry, multiply the corresponding row sum by the corresponding column sum, and then divide by the overall sum of peppers.

For example, for small red peppers, multiply the row sum (the 'red' row, which has sum 160) by the column sum (the 'small' column, which has sum 110), and then divide by the total number of peppers, which is 500. This gives $\dfrac{160 \times 110}{500} = 35.2$.

Similarly, for medium red peppers, we work out $\dfrac{160 \times 220}{500} = 70.4$, and so on. If you do all of these calculations, you should get the following table of expected values (note that the sums are all still the same).

	Small	Medium	Large	Sum
Red	35.2	70.4	54.4	160
Green	50.6	101.2	78.2	230
Orange	13.2	26.4	20.4	60
Yellow	11	22	17	50
Sum	110	220	170	500

Note that we *do not* round the answers here, to make our calculation as exact as possible (even though 35.2 peppers doesn't make a lot of sense!)

Now we can perform the χ^2-test as usual, by comparing the differences. Remember that we are testing whether there is a difference from what we would expect: we would expect the χ^2-test statistic to be 0 if they matched the predicted values exactly, and so we shall test how significantly different from 0 the actual χ^2-test statistic is.

Hypotheses
As discussed above, the null hypothesis is that there is no significant difference from the expected values, and so we have:

Null hypothesis: $\chi^2 = 0$
Alternative hypothesis: $\chi^2 > 0$

Test statistic
We need to work out the χ^2-test statistic, which is obtained, remember, by adding up all the differences squared divided by the expected values. The χ^2-test statistic is therefore as follows, where we take each value in the tables in turn, comparing the observed values with the expected values:

$\dfrac{(30 - 35.2)^2}{35.2} + \dfrac{(70 - 70.4)^2}{70.4} + \dots$ and so on, for all 12 values in the table (each possibility of red, green, orange, yellow, and small, medium, large). If you do this right, you should get a final answer of 5.447 (to 3 decimal places). This is our χ^2-test statistic.

Critical value
There are 4 rows and 3 columns, and so the number of degrees of freedom is $(4 - 1) \times (3 - 1) = 3 \times 2 = 6$, as discussed before.

We shall test at the 5% level: the corresponding test statistic (from Table D) for 6 degrees of freedom is 12.592.

Conclusion
The test statistic, 5.447, is less than the critical value of 12.592, and so there is not enough evidence at the 5% level to reject the null hypothesis. We therefore conclude that there is insufficient evidence to say there is any connection between colour and size.

● More on the χ^2-test

You can use the χ^2-test for any number of criteria. The only problem is that it's rather hard to write them down, but a computer can cope with it!

The test is not perfect, especially when there are small values involved. If there are a few values less than about 5 in the data, then it can be somewhat misleading, and it is very sensitive to values as small as 1. Really, you should try and combine classes together, if they have very small values, to get a more accurate answer.

When there is only one degree of freedom (which happens if there is one criterion taking two possible values, such as pass/fail, or two criteria, both taking two possible values, so the degrees of freedom are $(2 - 1) \times (2 - 1) = 1 \times 1 = 1$) then the χ^2-test is especially inaccurate. Something called *Yates' correction* deals with this. Essentially we subtract 0.5 from the absolute value (that is, the value ignoring negatives) of the differences. You probably won't need to use this, but I'm mentioning it so that you know about it if you do come across it.

There is always more to learn. You will never be taught everything there is to know about a subject. If you like it, and want to find out more, then go ahead and explore, and see what you can find. One of the most important, and most fun, aspects of study is taking your own responsibility to explore a topic further. You can surprise yourself by what you might learn, and you might surprise your lecturers too (they don't know everything!)

Summary

Although it has a rather strange name, the χ^2-test is a powerful, very widely used concept. Essentially, all we are asking here, is 'Is this data what we expected?' If we test it, and it is far away from what we expected, then surely we should investigate; it may be that nothing is wrong, but at least it has been flagged for us to look at. With our exams example, it might have been that the students were simply very good, and the exam was perfectly fair, but the analysis has indicated that it's different from the norm, and we should at least investigate. This sort of test ('Does what we got match what we expect?') is important in a huge range of subject areas.

 Exercises

1 In a packet of 100 sweets, it is intended that red, green, yellow, orange and black sweets would be spread evenly (so 20 of each), but it is accepted by customers that there may be some variation in this. A batch of sweets contains the following numbers. Perform a χ^2-test at the 5% level so see whether this batch of sweets is acceptable to be sold.

Colour	Red	Green	Yellow	Orange	Black
Number	13	26	19	14	28

2 For apples supplied to a supermarket to be sold at as a 'value range', it would be expected that 10% are below standard, 70% are average standard, and 20% are above standard. 200 apples from a huge consignment are tested to see whether they match this, and the following results are obtained. If they are declared unacceptable (they do not reasonably match this 10:70:20 ratio), then the batch is rejected and sent back. Use a χ^2-test at the 1% level to test whether this batch is acceptable or not.

Standard	Below standard	Standard	Above standard
Number	33	147	20

3 The following table summarises the performance of a group of 400 students at a university. Their grade is classified as A, B, C, D, E or F (fail), and their attendance at lectures is classified as poor, average or good. Add columns for the sum of each row and column, and then use a χ^2-test at the 1% level to see whether there is a strong relationship between the attendance and the grade obtained.

	A	B	C	D	E	F
Poor	8	10	12	24	16	30
Average	38	68	52	24	12	6
Good	54	82	36	16	12	0

4 As with other formulae we have considered, there is an alternative formulation of the χ^2-test statistic, which is to calculate $\chi^2 = \left| \sum_{i=1}^{k} \dfrac{O_i^2}{E_i} \right| - N$, where N is the total number of data values.

Check that this formula gives the same answer for the previous three questions.

4 | *F*-tests

The use of *F*-tests for comparison of variances

As well as comparing means, or checking to see whether values fit what we expect, it is also sometimes useful to compare the spread of data. Here we shall introduce a test (the *F*-test) that compares variances.

Key topics
- *F*-test

Key terms
comparison of variances *F*-test *F*-test statistic

Comparing means is very important, but it is also important to remember that there are two fundamental aspects to data: not only their average (e.g. their mean) but also their spread.

● Differences in variances

A good example of why we might need to compare variances is in scientific testing. Two methods might have the same mean, but wildly different variances, which means that one method's results are much more spread out.

For example, suppose we expect a test to return a mean of 6. A test that gives results 2, 4, 6, 8, 10 has the same mean (exactly 6) as a test that gives results 5.8, 5.9, 6.0, 6.1, 6.2, but clearly the second test appears to be more precise and doesn't give readings far away from the real answer: the second test has a much lower variance.

So, as well as comparing means, we should also have a test that allows us to compare variances. This is where the *F*-test comes in. The F has no real meaning; it is generally accepted that it was named after a man called Fisher, who worked heavily in this area in the 1920s.

Why should we compare variances and not standard deviations? We could, but using variances avoids the need for square roots, and so it seems to have evolved more naturally to use them rather than standard deviations.

● The F-test statistic

Performing an F-test is quite straightforward: you simply divide the variances together and compare your answer with a critical value obtained from a table. The main thing to note is that the value should be more than 1, and so we put the larger value on the top of the division.

If one sample has standard deviation s_1 and another sample has standard deviation s_2, then the sample variances are s_1^2 and s_2^2 (the variance is the square of the standard deviation). Then we can define the F-test statistic as follows:

The F-test statistic is given by $\dfrac{s_1^2}{s_2^2}$, where $s_1^2 > s_2^2$

Remember that we choose s_1 and s_2 the right way round, so that this answer will be more than 1 (so the largest value will be s_1 and go on the top of the fraction).

Let's look at an example of this.

● F-test: an example

Two groups of chemistry students try to perform a test as accurately as possible, so the variance/standard deviation ought to be low. The first group, of 16 students, have a standard deviation of 1.2. The second group, of 12 students, have a standard deviation of 1.4. Is there a significant difference between the precision of the two groups?

The null hypothesis is that the variances should be the same, and the difference is just down to experimental error, whereas the alternative hypothesis is that there is a genuine difference between the two groups (perhaps indicating a problem with the equipment).

We want to divide one variance by the other, but it is important that the bigger number goes on the top of the fraction, so we work out $\dfrac{(1.4)^2}{(1.2)^2} = 1.361$ (to 3 decimal places).

Let us test at the 5% level. We shall use Table E1 in the Statistical Tables section, which gives the critical values for different sample sizes at the 5% level. We have sample sizes of 12 and 16. Remember the order: the one corresponding to 1.4 (which is 12) comes first, and the other (corresponding to 1.2) comes second. Looking up 12 and 16 in this table gives 2.60.

Since our test statistic 1.361 is less than this, we do not have enough evidence at the 5% level to reject the null hypothesis, and so do not have evidence that the groups are genuinely different in their variances.

 ## Summary

This chapter is very brief, but there is not much more to say. To test whether the variances are significantly different, just divide them and compare with a value in a table; that is all there is to it.

> Every subject has its own difficulties, and some subjects seem easier than others. There isn't much to learn here, compared with some of the other topics we have discussed, but that doesn't mean it's any more or less important; it's natural that some things take longer to explain than others.

 ## Exercises

1 Two large groups of students take a test. Ten students are sampled from each group. Both groups have the same mean, but the standard deviation of the first group is only 2, while the standard deviation of the second group is 4. Is this evidence of a significant difference between the groups? Test at both the 5% and 1% levels. (Use Tables E1 and E2.)

2 A sample of 15 tests from one company shows a standard deviation of 2.1, and a sample of 12 tests from another company shows a standard deviation of 8.2. Is there evidence that the variances are significantly different? Test at both the 5% and 1% levels.

STATISTICAL TABLES

● Table A The normal distribution $N(0,1)$

This table gives the area of the normal distribution between 0 and the specified value. The specified values are given to 1 decimal place on the left-hand side; then you should read along to get the appropriate value for 2 decimal places. For example, for 1.26 go to the row labelled 1.2, then read across until you reach 0.06, so that you are at 1.26, and read off the value 0.3962.

x	0.00	0.01	0.02	0.03	0.04	0.05	0.06	0.07	0.08	0.09
0.0	0.0000	0.0040	0.0080	0.0120	0.0160	0.0199	0.0239	0.0279	0.0319	0.0359
0.1	0.0398	0.0438	0.0478	0.0517	0.0557	0.0596	0.0636	0.0675	0.0714	0.0753
0.2	0.0793	0.0832	0.0871	0.0910	0.0948	0.0987	0.1026	0.1064	0.1103	0.1141
0.3	0.1179	0.1217	0.1255	0.1293	0.1331	0.1368	0.1406	0.1443	0.1480	0.1517
0.4	0.1554	0.1591	0.1628	0.1664	0.1700	0.1736	0.1772	0.1808	0.1844	0.1879
0.5	0.1915	0.1950	0.1985	0.2019	0.2054	0.2088	0.2123	0.2157	0.2190	0.2224
0.6	0.2257	0.2291	0.2324	0.2357	0.2389	0.2422	0.2454	0.2486	0.2517	0.2549
0.7	0.2580	0.2611	0.2642	0.2673	0.2704	0.2734	0.2764	0.2794	0.2823	0.2852
0.8	0.2881	0.2910	0.2939	0.2967	0.2995	0.3023	0.3051	0.3078	0.3106	0.3133
0.9	0.3159	0.3186	0.3212	0.3238	0.3264	0.3289	0.3315	0.3340	0.3365	0.3389
1.0	0.3413	0.3438	0.3461	0.3485	0.3508	0.3531	0.3554	0.3577	0.3599	0.3621
1.1	0.3643	0.3665	0.3686	0.3708	0.3729	0.3749	0.3770	0.3790	0.3810	0.3830
1.2	0.3849	0.3869	0.3888	0.3907	0.3925	0.3944	0.3962	0.3980	0.3997	0.4015
1.3	0.4032	0.4049	0.4066	0.4082	0.4099	0.4115	0.4131	0.4147	0.4162	0.4177
1.4	0.4192	0.4207	0.4222	0.4236	0.4251	0.4265	0.4279	0.4292	0.4306	0.4319
1.5	0.4332	0.4345	0.4357	0.4370	0.4382	0.4394	0.4406	0.4418	0.4429	0.4441
1.6	0.4452	0.4463	0.4474	0.4484	0.4495	0.4505	0.4515	0.4525	0.4535	0.4545
1.7	0.4554	0.4564	0.4573	0.4582	0.4591	0.4599	0.4608	0.4616	0.4625	0.4633
1.8	0.4641	0.4649	0.4656	0.4664	0.4671	0.4678	0.4686	0.4693	0.4699	0.4706
1.9	0.4713	0.4719	0.4726	0.4732	0.4738	0.4744	0.4750	0.4756	0.4761	0.4767
2.0	0.4772	0.4778	0.4783	0.4788	0.4793	0.4798	0.4803	0.4808	0.4812	0.4817
2.1	0.4821	0.4826	0.4830	0.4834	0.4838	0.4842	0.4846	0.4850	0.4854	0.4857
2.2	0.4861	0.4864	0.4868	0.4871	0.4875	0.4878	0.4881	0.4884	0.4887	0.4890
2.3	0.4893	0.4896	0.4898	0.4901	0.4904	0.4906	0.4909	0.4911	0.4913	0.4916
2.4	0.4918	0.4920	0.4922	0.4925	0.4927	0.4929	0.4931	0.4932	0.4934	0.4936

(continued)

x	0.00	0.01	0.02	0.03	0.04	0.05	0.06	0.07	0.08	0.09
2.5	0.4938	0.4940	0.4941	0.4943	0.4945	0.4946	0.4948	0.4949	0.4951	0.4952
2.6	0.4953	0.4955	0.4956	0.4957	0.4959	0.4960	0.4961	0.4962	0.4963	0.4964
2.7	0.4965	0.4966	0.4967	0.4968	0.4969	0.4970	0.4971	0.4972	0.4973	0.4974
2.8	0.4974	0.4975	0.4976	0.4977	0.4977	0.4978	0.4979	0.4979	0.4980	0.4981
2.9	0.4981	0.4982	0.4982	0.4983	0.4984	0.4984	0.4985	0.4985	0.4986	0.4986
3.0	0.4987	0.4987	0.4987	0.4988	0.4988	0.4989	0.4989	0.4989	0.4990	0.4990
3.1	0.4990	0.4991	0.4991	0.4991	0.4992	0.4992	0.4992	0.4992	0.4993	0.4993
3.2	0.4993	0.4993	0.4994	0.4994	0.4994	0.4994	0.4994	0.4995	0.4995	0.4995
3.3	0.4995	0.4995	0.4995	0.4996	0.4996	0.4996	0.4996	0.4996	0.4996	0.4997
3.4	0.4997	0.4997	0.4997	0.4997	0.4997	0.4997	0.4997	0.4997	0.4997	0.4998
3.5	0.4998	0.4998	0.4998	0.4998	0.4998	0.4998	0.4998	0.4998	0.4998	0.4998

● Table B Critical values and confidence limits for the z-test

These tables give the critical values for various values of the z-test statistic for various percentages of confidence. The null hypothesis is rejected if the z-test statistic lies outside the critical range for the particular percentage.

One-tailed tests

One-tailed tests are used when you are testing whether one mean is larger than another. For a particular percentage, if z lies above the corresponding value, then you can reject the null hypothesis at this percentage value.

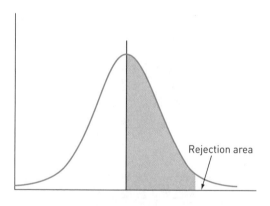

Percentage to test at	Value
99.5% (or 0.5% level)	2.5758
99% (or 1% level)	2.3263
98% (or 2% level)	2.0537
97.5% (or 2.5% level)	1.96
97% (or 3% level)	1.8801
96% (or 4% level)	1.7507
95% (or 5% level)	1.6449
90% (or 10% level)	1.2816
85% (or 15% level)	1.0364
80% (or 20% level)	0.8416
50% (or 50% level)	0

For a one-tailed test where you are testing whether one mean is smaller than another, the values can be read from this table simply by making them negative (and you then reject the null hypothesis if the z value lies below this value).

Two-tailed tests

Two-tailed tests are used when you are simply testing whether one mean is different from another. For a particular percentage, if z lies outside the corresponding range, then you can reject the null hypothesis at this percentage value.

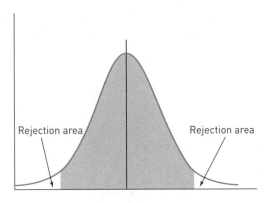

Percentage to test at	Range
99.5% (or 0.5% level)	−2.807 to 2.807
99% (or 1% level)	−2.5758 to 2.5758
98% (or 2% level)	−2.3263 to 2.3263
97.5% (or 2.5% level)	−2.2414 to 2.2414
97% (or 3% level)	−2.1701 to 2.1701
96% (or 4% level)	−2.0537 to 2.0537
95% (or 5% level)	−1.9600 to 1.9600
90% (or 10% level)	−1.6449 to 1.6449
85% (or 15% level)	−1.4395 to 1.4395
80% (or 20% level)	−1.2816 to 1.2816
50% (or 50% level)	−0.6745 to 0.6745

● Table C1 The one-tailed *t*-test

This table gives the one-tailed *t*-test values for various degrees of freedom and selected confidence levels; everything from 1 to 30 is given, and selected others (use 'Infinite' if you have a very large sample). Remember that the number of degrees of freedom is one less than the sample size. If you are testing for a negative value, then simply use the negative of the value given in this table.

Degrees of freedom ($n - 1$)	90% (or 10% level)	95% (or 5% level)	97.5% (or 2.5% level)	99% (or 1% level)
1	3.078	6.314	12.706	31.821
2	1.886	2.920	4.303	6.965
3	1.638	2.353	3.182	4.541
4	1.533	2.132	2.776	3.747
5	1.476	2.015	2.571	3.365
6	1.440	1.943	2.447	3.143
7	1.415	1.895	2.365	2.998
8	1.397	1.860	2.306	2.896
9	1.383	1.833	2.262	2.821
10	1.372	1.812	2.228	2.764
11	1.363	1.796	2.201	2.718
12	1.356	1.782	2.179	2.681
13	1.350	1.771	2.160	2.650
14	1.345	1.761	2.145	2.624
15	1.341	1.753	2.131	2.602
16	1.337	1.746	2.120	2.583
17	1.333	1.740	2.110	2.567
18	1.330	1.734	2.101	2.552
19	1.328	1.729	2.093	2.539
20	1.325	1.725	2.086	2.528
21	1.323	1.721	2.080	2.518
22	1.321	1.717	2.074	2.508
23	1.319	1.714	2.069	2.500
24	1.318	1.711	2.064	2.492

▶

(continued)

Degrees of freedom $(n-1)$	90% (or 10% level)	95% (or 5% level)	97.5% (or 2.5% level)	99% (or 1% level)
25	1.316	1.708	2.060	2.485
26	1.315	1.706	2.056	2.479
27	1.314	1.703	2.052	2.473
28	1.313	1.701	2.048	2.467
29	1.311	1.699	2.045	2.462
30	1.310	1.697	2.042	2.457
40	1.303	1.684	2.021	2.423
60	1.296	1.671	2.000	2.390
120	1.289	1.658	1.980	2.358
Infinite	1.282	1.645	1.960	2.326

● Table C2 The two-tailed *t*-test

This table gives the two-tailed *t*-test values for various degrees of freedom and selected confidence levels. Everything from 1 to 30 is given, and selected others (use 'Infinite' if you have a very large sample). Remember that the number of degrees of freedom is one less than the sample size. The confidence range is between the negative value and the positive value, so for example if the figure is 3.182, the range is from −3.182 to 3.182.

Degrees of freedom ($n - 1$)	90% (or 10% level)	95% (or 5% level)	97.5% (or 2.5% level)	99% (or 1% level)
1	6.314	12.706	25.452	63.657
2	2.920	4.303	6.205	9.925
3	2.353	3.182	4.177	5.841
4	2.132	2.776	3.495	4.604
5	2.015	2.571	3.163	4.032
6	1.943	2.447	2.969	3.707
7	1.895	2.365	2.841	3.499
8	1.860	2.306	2.752	3.355
9	1.833	2.262	2.685	3.250
10	1.812	2.228	2.634	3.169
11	1.796	2.201	2.593	3.106
12	1.782	2.179	2.560	3.055
13	1.771	2.160	2.533	3.012
14	1.761	2.145	2.510	2.977
15	1.753	2.131	2.490	2.947
16	1.746	2.120	2.473	2.921
17	1.740	2.110	2.458	2.898
18	1.734	2.101	2.445	2.878
19	1.729	2.093	2.433	2.861
20	1.725	2.086	2.423	2.845
21	1.721	2.080	2.414	2.831
22	1.717	2.074	2.405	2.819
23	1.714	2.069	2.398	2.807

(continued)

Degrees of freedom $(n - 1)$	90% (or 10% level)	95% (or 5% level)	97.5% (or 2.5% level)	99% (or 1% level)
24	1.711	2.064	2.391	2.797
25	1.708	2.060	2.385	2.787
26	1.706	2.056	2.379	2.779
27	1.703	2.052	2.373	2.771
28	1.701	2.048	2.368	2.763
29	1.699	2.045	2.364	2.756
30	1.697	2.042	2.360	2.750
40	1.684	2.021	2.329	2.704
60	1.671	2.000	2.299	2.660
120	1.658	1.980	2.270	2.617
Infinite	1.645	1.960	2.241	2.576

● Table D Critical values for the χ^2 distribution

This table gives the critical values for various degrees of freedom, and various confidence levels, for the χ^2 distribution. With one criterion, the degrees of freedom is one less than the number of classes; with two criteria with m and n classes, the number of degrees of freedom is $(m - 1)(n - 1)$.

Degrees of freedom	90% (or 10% level)	95% (or 5% level)	97.5% (or 2.5% level)	99% (or 1% level)
1	2.706	3.841	5.024	6.635
2	4.605	5.991	7.378	9.210
3	6.251	7.815	9.348	11.345
4	7.779	9.488	11.143	13.277
5	9.236	11.070	12.833	15.086
6	10.645	12.592	14.449	16.812
7	12.017	14.067	16.013	18.475
8	13.362	15.507	17.535	20.090
9	14.684	16.919	19.023	21.666
10	15.987	18.307	20.483	23.209
11	17.275	19.675	21.920	24.725
12	18.549	21.026	23.337	26.217
13	19.812	22.362	24.736	27.688
14	21.064	23.685	26.119	29.141
15	22.307	24.996	27.488	30.578
16	23.542	26.296	28.845	32.000
17	24.769	27.587	30.191	33.409
18	25.989	28.869	31.526	34.805
19	27.204	30.144	32.852	36.191
20	28.412	31.410	34.170	37.566
21	29.615	32.671	35.479	38.932
22	30.813	33.924	36.781	40.289
23	32.007	35.172	38.076	41.638
24	33.196	36.415	39.364	42.980
25	34.382	37.652	40.646	44.314

(continued)

Degrees of freedom	90% (or 10% level)	95% (or 5% level)	97.5% (or 2.5% level)	99% (or 1% level)
26	35.563	38.885	41.923	45.642
27	36.741	40.113	43.195	46.963
28	37.916	41.337	44.461	48.278
29	39.087	42.557	45.722	49.588
30	40.256	43.773	46.979	50.892
40	51.805	55.758	59.342	63.691
50	63.167	67.505	71.420	76.154
60	74.397	79.082	83.298	88.379
70	85.527	90.531	95.023	100.425
80	96.578	101.879	106.629	112.329
90	107.565	113.145	118.136	124.116
100	118.498	124.342	129.561	135.807
110	129.385	135.480	140.917	147.414
120	140.233	146.567	152.211	158.950

● Table E1 Critical values for the *F*-test (5%)

The values must be chosen in the right order, so that the *F*-test statistic is greater than 1. Then look up the appropriate value from these tables, depending on your sample sizes.

	1	2	3	4	5	6	7	8	9	10
1	161.45	199.50	215.71	224.58	230.16	233.99	236.77	238.88	240.54	241.88
2	18.51	19.00	19.16	19.25	19.30	19.33	19.35	19.37	19.38	19.40
3	10.13	9.55	9.28	9.12	9.01	8.94	8.89	8.85	8.81	8.79
4	7.71	6.94	6.59	6.39	6.26	6.16	6.09	6.04	6.00	5.96
5	6.61	5.79	5.41	5.19	5.05	4.95	4.88	4.82	4.77	4.74
6	5.99	5.14	4.76	4.53	4.39	4.28	4.21	4.15	4.10	4.06
7	5.59	4.74	4.35	4.12	3.97	3.87	3.79	3.73	3.68	3.64
8	5.32	4.46	4.07	3.84	3.69	3.58	3.50	3.44	3.39	3.35
9	5.12	4.26	3.86	3.63	3.48	3.37	3.29	3.23	3.18	3.14
10	4.96	4.10	3.71	3.48	3.33	3.22	3.14	3.07	3.02	2.98
11	4.84	3.98	3.59	3.36	3.20	3.09	3.01	2.95	2.90	2.85
12	4.75	3.89	3.49	3.26	3.11	3.00	2.91	2.85	2.80	2.75
13	4.67	3.81	3.41	3.18	3.03	2.92	2.83	2.77	2.71	2.67
14	4.60	3.74	3.34	3.11	2.96	2.85	2.76	2.70	2.65	2.60
15	4.54	3.68	3.29	3.06	2.90	2.79	2.71	2.64	2.59	2.54
16	4.49	3.63	3.24	3.01	2.85	2.74	2.66	2.59	2.54	2.49
17	4.45	3.59	3.20	2.96	2.81	2.70	2.61	2.55	2.49	2.45
18	4.41	3.55	3.16	2.93	2.77	2.66	2.58	2.51	2.46	2.41
19	4.38	3.52	3.13	2.90	2.74	2.63	2.54	2.48	2.42	2.38
20	4.35	3.49	3.10	2.87	2.71	2.60	2.51	2.45	2.39	2.35

	11	12	13	14	15	16	17	18	19	20
1	242.98	243.91	244.69	245.36	245.95	246.46	246.92	247.32	247.69	248.01
2	19.40	19.41	19.42	19.42	19.43	19.43	19.44	19.44	19.44	19.45
3	8.76	8.74	8.73	8.71	8.70	8.69	8.68	8.67	8.67	8.66
4	5.94	5.91	5.89	5.87	5.86	5.84	5.83	5.82	5.81	5.80
5	4.70	4.68	4.66	4.64	4.62	4.60	4.59	4.58	4.57	4.56
6	4.03	4.00	3.98	3.96	3.94	3.92	3.91	3.90	3.88	3.87
7	3.60	3.57	3.55	3.53	3.51	3.49	3.48	3.47	3.46	3.44
8	3.31	3.28	3.26	3.24	3.22	3.20	3.19	3.17	3.16	3.15
9	3.10	3.07	3.05	3.03	3.01	2.99	2.97	2.96	2.95	2.94
10	2.94	2.91	2.89	2.86	2.85	2.83	2.81	2.80	2.79	2.77
11	2.82	2.79	2.76	2.74	2.72	2.70	2.69	2.67	2.66	2.65
12	2.72	2.69	2.66	2.64	2.62	2.60	2.58	2.57	2.56	2.54
13	2.63	2.60	2.58	2.55	2.53	2.51	2.50	2.48	2.47	2.46
14	2.57	2.53	2.51	2.48	2.46	2.44	2.43	2.41	2.40	2.39
15	2.51	2.48	2.45	2.42	2.40	2.38	2.37	2.35	2.34	2.33
16	2.46	2.42	2.40	2.37	2.35	2.33	2.32	2.30	2.29	2.28
17	2.41	2.38	2.35	2.33	2.31	2.29	2.27	2.26	2.24	2.23
18	2.37	2.34	2.31	2.29	2.27	2.25	2.23	2.22	2.20	2.19
19	2.34	2.31	2.28	2.26	2.23	2.21	2.20	2.18	2.17	2.16
20	2.31	2.28	2.25	2.22	2.20	2.18	2.17	2.15	2.14	2.12

The values must be chosen in the right order, so that the F-test statistic is greater than 1. Then look up the appropriate value from these tables, depending on your sample sizes.

	1	2	3	4	5	6	7	8	9	10
1	4052	4999	5403	5625	5764	5859	5928	5981	6022	6056
2	98.50	99.00	99.17	99.25	99.30	99.33	99.36	99.37	99.39	99.40
3	34.12	30.82	29.46	28.71	28.24	27.91	27.67	27.49	27.35	27.23
4	21.20	18.00	16.69	15.98	15.52	15.21	14.98	14.80	14.66	14.55
5	16.26	13.27	12.06	11.39	10.97	10.67	10.46	10.29	10.16	10.05
6	13.75	10.92	9.78	9.15	8.75	8.47	8.26	8.10	7.98	7.87
7	12.25	9.55	8.45	7.85	7.46	7.19	6.99	6.84	6.72	6.62
8	11.26	8.65	7.59	7.01	6.63	6.37	6.18	6.03	5.91	5.81
9	10.56	8.02	6.99	6.42	6.06	5.80	5.61	5.47	5.35	5.26
10	10.04	7.56	6.55	5.99	5.64	5.39	5.20	5.06	4.94	4.85
11	9.65	7.21	6.22	5.67	5.32	5.07	4.89	4.74	4.63	4.54
12	9.33	6.93	5.95	5.41	5.06	4.82	4.64	4.50	4.39	4.30
13	9.07	6.70	5.74	5.21	4.86	4.62	4.44	4.30	4.19	4.10
14	8.86	6.51	5.56	5.04	4.69	4.46	4.28	4.14	4.03	3.94
15	8.68	6.36	5.42	4.89	4.56	4.32	4.14	4.00	3.89	3.80
16	8.53	6.23	5.29	4.77	4.44	4.20	4.03	3.89	3.78	3.69
17	8.40	6.11	5.18	4.67	4.34	4.10	3.93	3.79	3.68	3.59
18	8.29	6.01	5.09	4.58	4.25	4.01	3.84	3.71	3.60	3.51
19	8.18	5.93	5.01	4.50	4.17	3.94	3.77	3.63	3.52	3.43
20	8.10	5.85	4.94	4.43	4.10	3.87	3.70	3.56	3.46	3.37

	11	12	13	14	15	16	17	18	19	20
1	6083	6106	6126	6143	6157	6170	6181	6192	6201	6209
2	99.41	99.42	99.42	99.43	99.43	99.44	99.44	99.44	99.45	99.45
3	27.13	27.05	26.98	26.92	26.87	26.83	26.79	26.75	26.72	26.69
4	14.45	14.37	14.31	14.25	14.20	14.15	14.11	14.08	14.05	14.02
5	9.96	9.89	9.82	9.77	9.72	9.68	9.64	9.61	9.58	9.55
6	7.79	7.72	7.66	7.60	7.56	7.52	7.48	7.45	7.42	7.40
7	6.54	6.47	6.41	6.36	6.31	6.28	6.24	6.21	6.18	6.16
8	5.73	5.67	5.61	5.56	5.52	5.48	5.44	5.41	5.38	5.36
9	5.18	5.11	5.05	5.01	4.96	4.92	4.89	4.86	4.83	4.81
10	4.77	4.71	4.65	4.60	4.56	4.52	4.49	4.46	4.43	4.41
11	4.46	4.40	4.34	4.29	4.25	4.21	4.18	4.15	4.12	4.10
12	4.22	4.16	4.10	4.05	4.01	3.97	3.94	3.91	3.88	3.86
13	4.02	3.96	3.91	3.86	3.82	3.78	3.75	3.72	3.69	3.66
14	3.86	3.80	3.75	3.70	3.66	3.62	3.59	3.56	3.53	3.51
15	3.73	3.67	3.61	3.56	3.52	3.49	3.45	3.42	3.40	3.37
16	3.62	3.55	3.50	3.45	3.41	3.37	3.34	3.31	3.28	3.26
17	3.52	3.46	3.40	3.35	3.31	3.27	3.24	3.21	3.19	3.16
18	3.43	3.37	3.32	3.27	3.23	3.19	3.16	3.13	3.10	3.08
19	3.36	3.30	3.24	3.19	3.15	3.12	3.08	3.05	3.03	3.00
20	3.29	3.23	3.18	3.13	3.09	3.05	3.02	2.99	2.96	2.94

SUMMARY, GLOSSARY AND APPENDICES

Summary and further work

Data is everywhere. Especially in this digital age, there is an incomprehensible amount of data, which needs to be analysed. Think of all the people and companies out there, all analysing their sales figures, or their customer opinions, making decisions whether to switch suppliers, whether they are performing better than their competitors, and so much more.

Statistics is necessary for all of this. The aim of this book is to make you see why statistics is so important – to give you some ideas as to why it needs to be used.

However, this book is only an introduction. Statistics is a huge field, and there is so much more that you could find out about and learn. We have talked about some of the tests you can do – some of the measures of averages, and so on – but this only scratches the surface.

If you find topics interesting, then go ahead and find out more. Hopefully, you have gained some level of interest, and understand why statistics is relevant to your studies, but do explore further. If you find anything interesting, let me know, and I can put relevant information on the website.

It only remains for me to wish you good luck with whatever course you are studying; I hope you will find some of the material here useful in the future. Good luck!

Glossary

The following is a list of the statistical terms used in the book:

Accept for a particular hypothesis, to make a conclusion that there is evidence that the hypothesis is true.

Accumulator a bet on more than one event, where you have to win all the events to win the bet overall.

Alternative hypothesis the claim that we are testing to see if we can accept it as true, rather than the default of the null hypothesis. Usually denoted by H_1.

Average any method of working out the 'typical' value of a set of data, which is often the mean.

Bar chart a graph used to display data that falls into categories.

Bayes' theorem the conditional probability theorem that states
$$P(A|B) = \frac{P(B|A) \times P(A)}{P(B)}.$$

Bimodal a set of data that has two modes (there are two values that appear the most number of times).

Binomial distribution a discrete probability distribution that models the situation when we have many repeated yes/no events.

Chi-squared test a test as to how well observed data fits expectations.

Chi-squared test statistic the value used in a χ^2 test, which is given by
$\chi^2 = \sum_{i=1}^{k} \frac{(O_i - E_i)^2}{E_i}$, where k is the number of classes, O_i are the observed frequencies, and E_i are the expected frequencies.

Combination the number of ways to choose r things from a set of n things, where the order is not important. It is often written as ^nC_r, and is given by
$$\frac{n!}{r!(n - r)!}.$$

Complement the 'opposite' of a probabilistic event: for example, the complement of 'rolling a 6' is 'not rolling a 6'.

Conditional probability calculating probabilities where you already have knowledge that certain things have already happened. The conditional probability formula for A, where B has already happened, is given by
$$P(A|B) = \frac{P(A \cap B)}{P(B)}.$$

Confidence interval for a particular percentage, a range of values where the probability that a value lies in that range is that percentage: so, for example, a 95% confidence limit means that a value will lie in that range with 95% probability.

Contingency table a table presenting data classified with two criteria (such as colour and size).

Continuous data data that can take any value in a range, such as the speed of a car.

Correlation a relationship between two things. If the values of one thing depend strongly on another, then you would say there as a strong correlation; if they depend only weakly, then you would say that there is a weak correlation.

Critical value the endpoint or 'edge' of a confidence interval, which we use to compare a test statistic with.

Cumulative frequency the total number of values that occur up to, and including, a given category (so including all the previous categories).

Data a collection of values.

Datum the singular of data, so one particular value.

Degrees of freedom the number of free variables. For a sample of size n, this is $n - 1$; when there are two variables, m and n, it is $(m - 1)(n - 1)$.

Descriptive statistics presenting aspects of data (either visually or numerically).

Discrete data data that can take distinct, separate values, such as the roll of a die, which can take the specific values 1, 2, 3, 4, 5 or 6, and nothing else.

Disjoint another word for mutually exclusive.

e a special number in mathematics, approximately equal to 2.718281828.

Element one of the values in a set or list.

Event a particular occurrence, such as rolling a 6 with a roll of a die.

Expected frequencies the total number of data values falling in various categories that you would expect to occur.

Expected value the overall outcome you would expect in the long run if an event was repeated; obtained by multiplying together each possible outcome and its probability, and then adding them all together.

Expected value criterion deciding which option to take by comparing the expected values of the two options, and choosing the option with the higher value.

Factorial the number obtained by multiplying together all the positive integers up to the given number. For example, the factorial of 5, written as 5!, is 5! $= 5 \times 4 \times 3 \times 2 \times 1 = 120$.

False positive another phrase for a Type I error.

Frequency distribution placing data values into categories, and recording the number of values that are placed in each category.

F-test a test to compare two variances.

F-test statistic if one sample has standard deviation s_1 and another sample has standard deviation s_2, then the F-test statistic is $\dfrac{s_1^2}{s_2^2}$, where $s_1^2 > s_2^2$

Geometric mean an average obtained by multiplying all the values together, and then taking the nth root, where n is the number of values.

Gradient a measure of the slope of a straight line.

Harmonic mean an average obtained by adding together all the reciprocals of the values, and then dividing the total number of values by this sum.

Histogram another word for bar chart.

Hypothesis a proposal or statement that we wish to test to see whether it is true, such as 'This product works better than the old product.'

Hypothesis testing performing a statistical test on data to see whether a hypothesis about it is true.

Independent describes two events that have no connection between them, so that one does not affect the other.

Inferential statistics used to make conclusions about a set of data.

Intercept the value at which a line cuts the y-axis.

Interquartile range the difference between the upper and lower quartiles.

Interval a category of continuous values, such as the interval from 40 to 50, which contains every value between 40 and 50.

Kurtosis a measure as to how 'peaked' a probability distribution is: a large peak means it has high kurtosis, whereas if it is quite flat, it has low kurtosis.

Lambda the Greek letter usually used for the average number of events in a particular timescale with the Poisson distribution.

Laplace criterion says that, if you do not know the probability of events occurring, then assume they all have the same probability.

Line graph a graph of numerical data (usually taken over time), where straight lines are drawn to connect all the points.

Linear interpolation estimating where a value lies in a range by considering its expected position. For example, if we have sorted data and want the 45th value, and we know the 40th value is 100 and the 50th value is 200, then we can expect the 45th value to be halfway through this range, at 150.

List a collection of data, where values may be repeated.

Lower quartile the 25th percentile, so 25% of the data values are below the lower quartile.

Maximax criterion to make a decision by 'maximising your maximum gain', which generally means being as brave as possible and gambling in the hope of achieving something better.

Mean the average of a set of values, obtained by adding all the values and dividing by the total number of values.

Measure of central tendency another phrase for average.

Measure of dispersion another phrase for spread.

Measure of location another phrase for average.

Median The 'middle value' of a set of data, obtained by listing the values in ascending order and then choosing the middle value (if there are an odd number of values), or by taking the average of the two middle values (if there an even number of values).

Method of least squares to find the best-fit straight line for a set of data, by making the squares of the differences between the line and the actual values as small as possible.

Midpoint the point halfway along an interval.

Minimax criterion to make a decision by 'minimising your maximum loss', which generally means being as conservative as possible, and not taking any gambles, preferring to stay as you are.

Mode the most common value in a set of data.

Mutually exclusive when two events cannot both happen at the same time (such as rolling a 2 with a die, or rolling an odd number).

Negative correlation the situation where two things are related in the sense that, generally, as one thing gets bigger, the other thing gets smaller, such as the temperature compared with the altitude of an aeroplane: the higher the altitude, the colder it will be.

Normal distribution the most common continuous distribution; values can be obtained from tables.

$N(0, 1)$ distribution the normal distribution with mean 0 and standard deviation 1, from which all other normal distributions can be worked out.

Null hypothesis the claim (usually along the lines of 'nothing has changed') that we are testing, to see whether we can reject in favour of another claim, the alternative hypothesis. Usually denoted by H_0.

Observed frequencies the actual number of values falling into various categories from data that has been obtained.

Odds the chances given by bookmakers of an event (such as the winner of a horse race) happening.

One-tailed test a test where we are testing whether one thing is greater than something else (or less than something else), not just that they are different.

Paired t-test a t-test where the variables are paired together; often used in a 'before and after' scenario.

Path a particular route from start to finish through a probability tree.

Pearson's coefficient of skewness a measure of how 'skewed' a probability distribution is, given by 3(mean − median)/standard deviation.

Pearson's (product moment) correlation coefficient a measure of the correlation between two variables in a data set, given by the formula

$$\frac{\sum_{i=1}^{n}(x_i - \bar{x})(y_i - \bar{y})}{\sqrt{\sum_{i=1}^{n}(x_i - \bar{x})^2} \sqrt{\sum_{i=1}^{n}(y_i - \bar{y})^2}}.$$

Percentage level the level of confidence we have in a test result: so, for example, if we can be 99% confident in our example, we are testing at the 1% level.

Percentile the nth percentile is the value such that $n\%$ of the data values lie below it.

Permutation the number of ways to choose r things from a set of n things, where the order is important; often written as ^nP_r, and given by $\dfrac{n!}{(n-r)!}$.

Pi notation the use of the Greek capital letter Π to represent the multiplication of the terms in a sequence.

Pie chart a graph drawn as a circle, showing data placed in categories, where the size of each section of the circle depends on the number of values in each category.

Poisson distribution the probability distribution that says that if, on average, something happens λ times in a time period, then the probability of it actually happening r times in a similar time period is $\dfrac{\lambda^r e^{-\lambda}}{r!}$.

Pooled estimate an estimate of the overall standard deviation, obtained by $s_p = \sqrt{\dfrac{s_1^2 + s_2^2}{2}}$ where s_1 and s_2 are the two standard deviations.

Population the entire set of data being considered.

Population standard deviation the square root of the population variance.

Population variance a measure of the spread of a sample, obtained by calculating all the differences from the mean, squaring them all, then adding up all these squares and finally dividing by n, where n is the size of the sample.

Positive correlation the situation where two things are related in the sense that, generally, as one thing gets bigger, so does the other thing, such as the height and weight of a sample of people.

Probability the chance of something happening, which ranges from 0 (impossible) to 1 (certain); given by dividing the number of ways in which it can occur by the total number of possibilities.

Probability distribution a pictorial or numerical description of the probability of each of the possible events occurring.

Range (1) the difference between the largest value and the smallest value in a set of data.

Range (2) another word for interval.

Ranking where values are given a rank (first, second, third, etc.) rather than their actual value.

Rare event an event that has no real 'opposite' in a practical sense, such as a person calling a telephone help desk: the opposite, that they didn't call, is true for virtually every person in the world.

Reciprocal the number obtained from a value by dividing 1 by the value: so, for example, the reciprocal of 10 is $\dfrac{1}{10}$.

Regression the act of creating the best-fit line for a set of data.

Regret criterion a method of making a decision where you base your decision on how much the possible regret will be if you had not made the decision you did.

Reject for a hypothesis, to make the conclusion that the evidence is that the hypothesis is untrue.

Residual the distance between a point on a graph, and a straight line approximating to the data.

Rounding approximating a decimal to a specified number of decimal places.

Sample a small set of the data taken from the entire population.

Sample standard deviation the square root of the sample variance.

Sample variance a measure of the spread of a sample, obtained by calculating all the differences from the mean, squaring them all, then adding up all these squares and finally dividing by $n - 1$, where n is the size of the sample.

Scatter graph a graph drawn for data, where no lines are drawn to connect the data together.

Sequence a list of values, usually with some pattern to them, such as 2, 4, 6, 8, 10.

Series the sum of a sequence, such as $2 + 4 + 6 + 8 + 10$ (which gives 30).

Set a collection of data.

Sigma notation the use of the Greek letter Σ to represent the sum in a series.

Simpson's paradox the seemingly counterintuitive scenario where two averages are individually better than two others, but the others have the better overall average when combined together.

Skewness a measure of how much a probability distribution falls to the left or right.

Spearman's rank correlation coefficient a measure of correlation of ranked data, given by

$$\rho = 1 - \left| \frac{6 \sum\limits_{i=1}^{n} d_i^2}{n(n-1)(n+1)} \right|, \text{ where } d_i = x_i - y_i.$$

Spread a measure of how spread out the data is: that is, how much the values differ from the average value.

Standard deviation see *Sample standard deviation*, or *Population standard deviation*.

Standard error for a sample with standard deviation σ and size n, the standard error is defined as $\frac{\sigma}{n}$, and measures how representative the sample is of the whole population.

Statistics the branch of mathematics that deals with the study of data.

Student's *t*-distribution the distribution used for the *t*-test.

Table a collection of numbers that can be used to look up critical values for a particular distribution or test.

Term an element of a set or list.

Test statistic a particular value in a test that will be compared with critical values to decide whether to reject the null hypothesis or not.

Tree a way of representing probability via a diagram.

***t*-test** a statistical test, used when the underlying standard deviation is not known.

***t*-test statistic** for a sample of size n, with mean $\bar{x}$ and sample standard deviation s, from a population with known mean μ, the t-test statistic is given by $t = \dfrac{\bar{x} - \mu}{s / \sqrt{n}}$.

Two-tailed test a test where we are testing whether two things are different (either greater than or less than).

Type I error rejecting the null hypothesis when actually it was true.

Type II error not rejecting the null hypothesis, when actually the alternative hypothesis was true.

Uniform distribution when all events are equally likely (such as rolling a die).

Unpaired *t*-test a t-test where there is no 'pairing' of the data values; often used to compare two samples.

Upper quartile the 75th percentile, so 75% of the data values are below the upper quartile.

Utility function an assignment of numbers to possible outcomes that represent how important they actually are to you as an individual.

Variance see *Sample variance*, or *Population variance*.

Yates' correction the method of subtracting 0.5 from all the differences when we have only one degree of freedom in a χ^2 test.

Yes/no event an event with only two possible outcomes, such as head/tail or right/wrong.

***z*-test** a statistical test, often used to compare two means.

***z*-test statistic (1)** if a population has mean μ and standard deviation σ, for a sample of size n and mean $\bar{x}$, then the $\bar{z}$-test statistic is given by $z = \dfrac{\bar{x} - \mu}{\sigma/\sqrt{n}}$. This value can then be compared with a critical value.

***z*-test statistic (2)** used to compare the means of two samples. If $\bar{x}_1$ is the mean of a sample of size n_1 from a normal distribution with mean μ_1 and standard deviation σ_1, and $\bar{x}_2$ is the mean of a sample of size n_2 from a normal distribution with mean μ_2 and standard deviation σ_2, then the z-test statistic is

$$z = \frac{(\bar{x}_1 - \bar{x}_2) - (\mu_1 - \mu_2)}{\sqrt{\dfrac{\sigma_1^{\,2}}{n_1} + \dfrac{\sigma_2^{\,2}}{n_2}}}.$$

Appendix 1 'Use of a calculator' test

You need to be able to obtain the right answers from your calculator. Whatever make and model you have, make sure that you can get the same answers as those given here, which are questions similar to things you will have to work out as you go through the book. If not, you need to find out how. It is absolutely vital that you can use your own calculator correctly. Repeat questions as many times as you need to until you can get full marks on this test.

Questions

1 $\dfrac{2 + 3 + 5 + 10}{4}$

2 $\sqrt[3]{2 \times 4 \times 8}$

3 $\dfrac{1}{\frac{1}{2} + \frac{1}{3} + \frac{1}{6}}$

4 $\dfrac{75}{100} \times 36$

5 $10 + \dfrac{(30 - 20)}{(40 - 20)} \times (50 - 40)$

6 $\sqrt{\dfrac{(4 - 2)^2 + (6 - 2)^2 + (5 - 2)^2}{3}}$

7 $\dfrac{3 \times 5 + 8 \times 15 + 5 \times 25}{16}$

8 $\dfrac{7!}{(7 - 3)!}$

9 $\dfrac{8!}{5!(8 - 5)!}$

10 $1^2 + 2^2 + 3^2 + 4^2 + 5^2$

11 $\dfrac{256}{\sqrt{45.4}\ \sqrt{2320}}$

12 $1 - \dfrac{6 \times (1^2 + (-1)^2 + 0^2}{3 \times 4 \times 2}$

13 $\dfrac{1}{6} \times \dfrac{2}{13} + \dfrac{1}{2} \times \dfrac{3}{5}$

14 $\dfrac{1/4}{1/8}$

15 $\dfrac{3^4\ e^{-3}}{4!}$

16 $\dfrac{6.2 - 5.9}{0.9 / \sqrt{30}}$

Solutions

Decimal answers are given to 3 decimal places where it is necessary to round.

1 5	**2** 4	**3** 1	**4** 27
5 15	**6** 3.109	**7** 16.25	**8** 210
9 56	**10** 55	**11** 0.789	**12** 0.5
13 0.326	**14** 2	**15** 0.168	**16** 1.826

Appendix 2 The Greek alphabet

Greek letters are used a lot in mathematics and statistics; I am giving a full list here, so that you know which letter is which when you do come across them.

Name	Lower case	Upper case
alpha	α	A
beta	β	B
gamma	γ	Γ
delta	δ	Δ
epsilon	ϵ	E
zeta	ζ	Z
eta	η	H
theta	θ	Θ
iota	ι	I
kappa	κ	K
lambda	λ	Λ
mu	μ	M
nu	ν	N
xi	ξ	Ξ
omicron	o	O
pi	π	Π
rho	ρ	P
sigma	σ	Σ
tau	τ	T
upsilon	υ	Y
phi	ϕ	Φ
chi	χ	X
psi	ψ	Ψ
omega	ω	Ω

Appendix 3 Some useful Excel commands

This is not a book to teach you how to use a software package; there are many books and tutorials that can do that. However, many of you will have Excel as part of Microsoft Office, and so I'll give here a list of some useful Excel commands relevant to the things we have talked about in the book.

I'm assuming you know how to use Excel at a basic level, so that if a command is listed as, for example, AVERAGE, you know that you have to type =AVERAGE(*values*) into a cell, where the values can either be specified in the brackets, or refer to a collection of cells.

For example, you can type =AVERAGE(1, 5, 3, 5, 2, 8, 1, 10, 5, 2) for specific values, or something like =AVERAGE(A1:A10) if the values are inside the cells A1 to A10.

If you don't know this, or how to do it, I recommend you first take a basic 'introduction to Excel' tutorial; there are many on the Internet.

An Excel spreadsheet illustrating these commands should appear on the book webpage.

Note also that Excel can draw lots of different types of graph; it is probably best to learn how to use the different types simply by experimenting.

Useful Excel statistical commands

The examples for these commands assume that where there is a collection of data, the data is contained in the cells A1 to A10 (and B1 to B10 where there are pairs of data), and so is referred to as A1:A10 and B1:B10. Of course, you will have your data in different cells, but this provides an illustration of the command.

AVERAGE gives the (arithmetic) mean of the data: e.g. =AVERAGE(A1:A10).

CHIDIST gives the chi-squared value for a given number of degrees of freedom: e.g. =CHIDIST(12, 4).

CHITEST performs a chi-squared test on two data sets (observed and expected): e.g. =CHITEST(A1:A10, B1:B10), giving a test statistic you can check in tables.

COMBIN gives the number of combinations $^{n}C_{r}$: e.g. =COMBIN(6, 4) gives $^{6}C_{4}$ (which is 15).

FACT gives the factorial of a number: e.g. =FACT(5) gives 5! = 120

FTEST performs an F-test on two data sets, giving a probability to check: e.g. =FTEST(A1:A10, B1:B10).

GEOMEAN gives the geometric mean of the data: e.g. =GEOMEAN(A1:A10).

HARMEAN gives the harmonic mean of the data: e.g. =HARMEAN(A1:A10).

INTERCEPT gives the intercept of the linear regression line for a set of x and y values: e.g. INTERCEPT(B1:B10, A1:A10) (note that the y values have to come first).

MAX gives the maximum value in the data: e.g. =MAX(A1:A10).

MEDIAN gives the median of the data: e.g. =MEDIAN(A1:A10).

MIN gives the minimum value in the data: e.g. =MIN(A1:A10).

MODE gives the mode of the data: e.g. =MODE(A1:A10).

NORMDIST gives the area under the normal distribution curve between $-\infty$ (where ∞ is the symbol for infinity) and the value in question. Needs TRUE at the end to give the cumulative answer, so an example is =NORMDIST(1, 0, 1, TRUE), which gives the answer 0.8413, which matches our table value 0.3413 if you remember that the Excel command is also including the whole of the 0.5 from the negative part of the curve, whereas our table doesn't.

NORMINV gives the confidence limit value for a specific value: e.g. =NORMINV(0.05, 0, 1) gives the 95% one-tailed confidence limit (1.6449).

PEARSON gives the Pearson correlation coefficient for two sets of data: e.g. =PEARSON(A1:A10, B1:B10).

PERCENTILE gives the nth percentile of the data: e.g. =PERCENTILE(A1:A10, 20%).

PERMUT gives the number of permutations ${}^{n}P_{r}$: e.g. =PERMUT(6, 4) gives ${}^{6}P_{4}$ (which is 360).

POISSON gives the probability of a certain number of events happening, given that they follow a Poisson distribution. You need to give the number you want, and the mean, and also specify whether you are dealing with 'cumulative probabilities' or not (everything up to a given value, or just that particular value). An example is POISSON(3, 2, FALSE), which gives the probability of a value of 3, given that the mean is 2 (e.g. the probability of 3 calls to a call centre in a minute if we expect on average 2 calls per minute).

QUARTILE gives the upper quartile, median, or lower quartile, depending on whether you specify 1, 2, or 3: e.g. =QUARTILE(A1:A10,1) gives the lower quartile, =QUARTILE(A1:A10,2) gives the median, and =QUARTILE(A1:A10,3) gives the upper quartile.

ROUND rounds a number to a certain number of decimal places: e.g. =ROUND(3.1415, 3) rounds to 3 decimal places and gives 3.142.

SLOPE gives the gradient of the linear regression line for a set of x and y values: e.g. SLOPE(B1:B10, A1:A10) (note that the y values have to come first).

STDEV gives the sample standard deviation: e.g. =STDEV(A1:A10).

STDEVP gives the population standard deviation: e.g. =STDEVP(A1:A10).

SUM gives the total sum of the data (all the values added together): e.g. =SUM(A1:A10).

TRUNC truncates a number (so ignores any decimal part): e.g. =TRUNC(3.82) gives 3.

TDIST gives the t-test variable for a particular value, with a given number of degrees of freedom and a number of tails: e.g. TDIST(1.8, 9, 1) gives the one-tailed t-test value for 1.8, with 9 degrees of freedom.

TTEST performs a t-test on two sets of data. You have to specify whether you want a one-tailed or two-tailed test, and whether the data is paired or not. See the Excel Help for further assistance, but as an example, =TTEST(A1:A10, B1:B10, 1, 1) gives a one-tailed paired t-test value.

VAR gives the sample variance of the data: e.g. =VAR(A1:A10).

VARP gives the population variance of the data: e.g. =VARP(A1:A10).

ZTEST gives the probability value of a z-test: e.g. =ZTEST(A1:A10, 7, 0.35) gives the probability that the test data mean is genuinely greater than the known mean, if the known mean was 7 and standard deviation was 0.35. See the Excel Help for further information.

Solutions to exercises

Note: These are only the final answers. For fully worked solutions, plus additional questions, please see the website.

Chapter 1 Introduction to statistics and data

2

20 mph to <30 mph: 11   30 mph to <40 mph: 18
40 mph to <50 mph: 3   50 mph or more: 2

3

First match: Player X has an average of 30 and Player Y has an average of 29, so Player Y has the better average.

Second match: Player X has an average of 36 and Player Y has an average of 35, so Player Y has the better average.

Overall: Player X has an average of 33 and Player Y has an average of 34, so Player X has the better average.

This seems strange; Player Y had the better average in both games, but overall Player X has the better average.

Chapter 2 Presentation of data

These are only possible graphs – you may choose different scales, axis labels, etc, so don't worry if yours looks a little different!

1

(a)

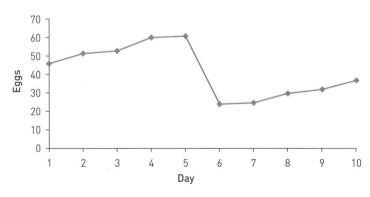

(b)

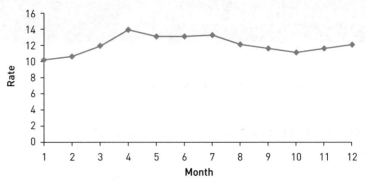

2 (a) There seems to be a positive correlation here.

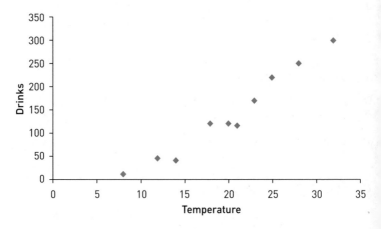

(b) There does not appear to be a correlation with batteries.

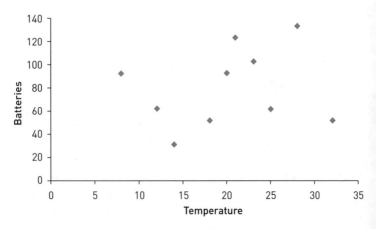

3 You should get something like the following, but again you may have used different scales, labels, etc.

(a)

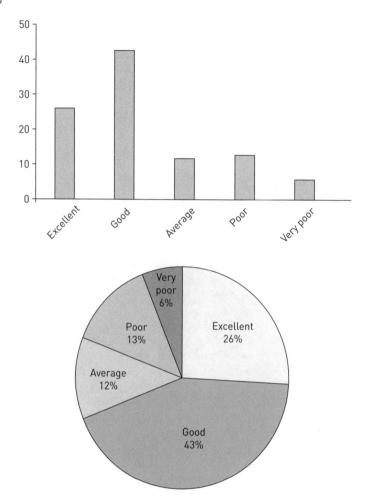

(b)

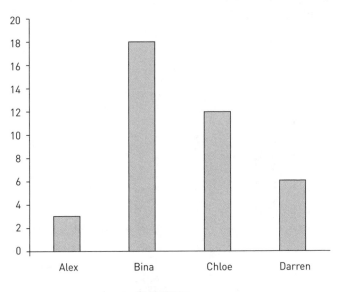

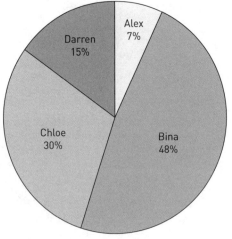

Chapter 3 Averages

1	(a) 7	(b) 3.833	(c) 5.5	(d) 3
2	(a) 5	(b) 7	(c) 7	(d) 5
3	(a) 2	(b) 1 and 4	(c) 3	(d) c
4	(a) 4	(b) 3.915	(c) 2.213	(d) 4
5	(a) 1.333	(b) 13.333	(c) 1.636	(d) 6

Chapter 4 Cumulative frequencies and percentiles

1

(a)

Speed	Number of cars	Cumulative frequency
>0 to ≤10 mph	2	2
>10 to ≤20 mph	3	5
>20 to ≤30 mph	5	10
>30 to ≤40 mph	28	38
>40 to ≤50 mph	56	94
>50 to ≤60 mph	16	110
>60 to ≤70 mph	9	119
>70 to ≤80 mph	1	120

(b)

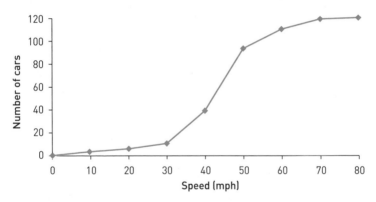

(c) Lower quartile = 37.412, median = 43.929, upper quartile = 49.286.

2

(a)

Score	Number	Cumulative frequency
>0 to ≤10	8	8
>10 to ≤20	38	46
>20 to ≤30	17	63
>30 to ≤40	5	68
>40 to ≤50	12	80
>50 to ≤60	20	100

(b)

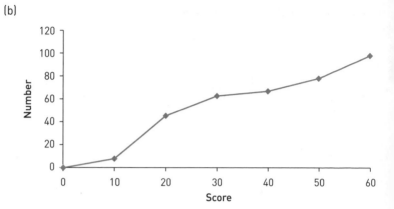

(c) Lower quartile = 14.474, median = 22.353, upper quartile = 45.833.

3

(a)

Mark	Number of students	Cumulative frequency
>0 to ≤10	4	4
>10 to ≤20	12	16
>20 to ≤30	14	30
>30 to ≤40	8	38
>40 to ≤50	52	90
>50 to ≤60	32	122
>60 to ≤70	28	150
>70 to ≤80	40	190
>80 to ≤90	8	198
>90 to ≤100	2	200

(b)

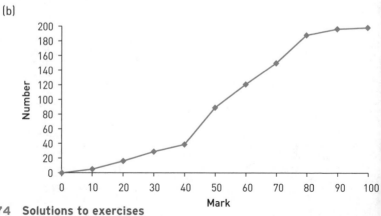

(c) Lower quartile = 42.308, median = 53.125, upper quartile = 70.

Chapter 5 Measures of dispersion

1 (a) 4 (b) 19 (c) 98 (d) 0

2 (a) 3.674 (b) 13.844 (c) 43.565 (d) 0

3 (a) 4.243 (b) 15.166 (c) 47.723 (d) 0

4 Group A has a mean of 40 and a standard deviation of 8.165, while Group B has a mean of 49 and a standard deviation of 34.809, so (a) Group B has a lower average, and (b) the marks are more spread out. (c) You can make an argument both ways as to which is the most successful: in Group A everyone passed, but in Group B those who passed did very well (the failures of 0 may well be because the students didn't turn up).

Chapter 6 Working with frequency distributions

1 Mean is 43.5.

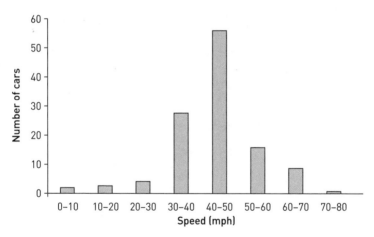

2 Mean is 28.5.

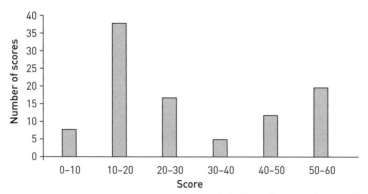

3 Mean is 53.1.

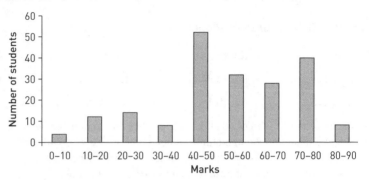

Chapter 7 Factorials, permutations and combinations

1 (a) 24   (b) 720   (c) 3,628,800   (d) 2   (e) 1   (f) 1

2 (a) $^6P_3 = 120$ and $^6C_3 = 20$ (b) $^8P_5 = 6720$ and $^8C_5 = 56$

   (c) $^{10}P_4 = 5040$ and $^{10}C_4 = 210$ (d) $^5P_2 = 20$ and $^5C_2 = 10$

   (e) $^8P_1 = 8$ and $^8C_1 = 8$ (f) $^5P_5 = 120$ and $^5C_5 = 1$

3 (a) 56   (b) 360   (c) 7.117×10^{12}

4 (a) 35   (b) 252   (c) 8568

5 (a) 7,893,600   (b) 40,320   (c) 126   (d) 13,983,816   (e) 88,200

Chapter 8 Sigma notation

1 (a) 21   (b) 36   (c) 10,000   (d) $\frac{1}{5}$ (e) 81   (f) 34

2 (a) $4 + 8 + 12 + 16 + 20 + 24 = 84$ (b) $3 + 5 + 7 + 9 = 24$

   (c) $1 + 2 + 4 + 8 + 16 = 31$ (d) $36 + 49 + 64 + 81 + 100 + 121 = 451$

   (e) $\frac{1}{2} + \frac{1}{4} + \frac{1}{8} + \frac{1}{16} + \ldots = 1$ (f) $1 + 2 + 3 + 4 + \ldots$, infinite sum

3 (a) Variance $= \frac{37}{3}$, standard deviation $= 3.512$ (3 decimal places)

   (b) Variance $= 10.64$, standard deviation $= 3.262$ (3 decimal places)

   (c) Variance $= 5.9375$, standard deviation $= 2.437$

   (d) Variance $= 0$, standard deviation $= 0$

4 (a) $\displaystyle\sum_{i=1}^{7} 4i$ (b) $\displaystyle\sum_{i=1}^{\infty} i$ (c) $\displaystyle\sum_{i=1}^{6} (2i - 1)$ (d) $\displaystyle\sum_{i=1}^{\infty} \frac{1}{2^{i-2}}$

Chapter 9 Correlation

1 (a) 0.942: a very strong positive correlation (the warmer the weather, the more visitors)

  (b) −0.726: a strong negative correlation (the longer worked, the less mistakes)

  (c) −0.409: a fairly weak negative correlation (taller players seem to score less, but not a hugely significant correlation)

2 0.568: a reasonably positive correlation (higher-placed teams have a better disciplinary record)

Chapter 10 Linear regression

1 (a) $y = 0.432x + 1.648$ (b) $y = -0.550x + 10.748$

  (c) $y = -18.646x + 40.190$

2 The answers with this formula match those in the chapter.

Chapter 11 An introduction to probability

1 (a) $\dfrac{1}{6}$ (b) $\dfrac{1}{3}$ (c) $\dfrac{1}{2}$

  (d) 1   (e) $\dfrac{5}{6}$ (f) 0

  (g) $\dfrac{2}{3}$ (h) $\dfrac{2}{3}$ (i) $\dfrac{5}{6}$

2 (a) $\dfrac{1}{26}$ (b) $\dfrac{2}{13}$ (c) $\dfrac{5}{26}$ (d) $\dfrac{21}{26}$ (e) $\dfrac{1}{2}$

3 The eight possibilities (using H for heads and T for tails) are (H, H, H), (H, H, T), (H, T, H), (H, T, T), (T, H, H), (T, H, T), (T, T, H), (T, T, T).

  (a) $\dfrac{1}{8}$ (b) $\dfrac{3}{8}$

4 You should switch; you have a $\dfrac{2}{3}$ chance of winning if you switch, and only a $\dfrac{1}{3}$ chance of winning if you don't switch.

Chapter 12 Multiple probabilities

1 (a) $\dfrac{1}{60}$ (b) $\dfrac{1}{4}$ (c) $\dfrac{1}{10}$ (d) $\dfrac{3}{4}$ (e) $\dfrac{1}{15}$ (f) $\dfrac{1}{10}$

2 (a) $\dfrac{3}{5}$ (b) $\dfrac{4}{5}$ (c) $\dfrac{1}{2}$ (d) 4.5

Chapter 13 Probability trees

1 The probability tree is as follows, where 'wt' stands for 'winning ticket' and 'lt' stands for 'losing ticket'. The probability of winning is $\frac{1}{2}$.

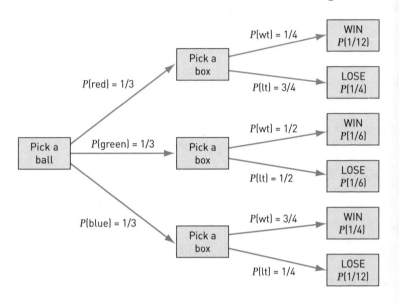

2 The probability tree for the first part is below; the probability of winning is $\frac{4}{9}$.

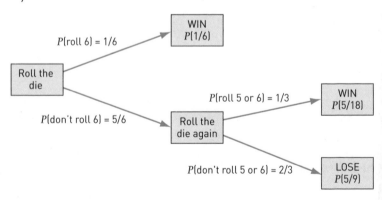

The second tree is below; the probability of winning is $\frac{29}{54}$.

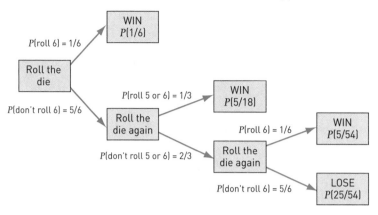

Chapter 14 Expected values and decision criteria

1 (a) $\frac{7}{2}$ or 3.5   (b) $\frac{10}{3}$ or 3.333 (3dp)   (c) $\frac{9}{2}$ or 4.5

  (d) 35   (e) 4

2 £4.75

3 Maximax, regret and expected value criteria: gamble and choose another box. Minimax and utility function: play safe and stay with your box.

4 (a) £25   (b) £75   (c) £50.50   (d) £20   (e) £24,310

5 £250,965

Chapter 15 Conditional probability

1 (a) $\frac{3}{4}$ (b) $\frac{1}{2}$

2 (a) $\frac{1}{2}$ (b) $\frac{1}{2}$ (c) 0

1 The graph looks like:

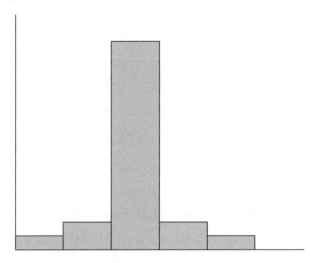

This graph has no skew (it is symmetrical), but has very high kurtosis (a very high peak).

2 The graph looks like:

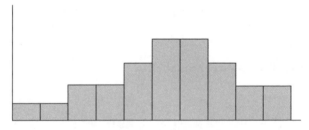

This graph is slightly skewed to the left (most of the possible values are to the left of the peak) and has quite low kurtosis; there isn't a very strong peak.

3 The graph looks like:

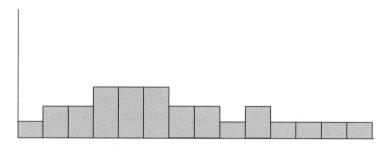

This graph is slightly skewed to the right (most of the possible values are to the right of the peak) and has very low kurtosis; it is very flat with barely a peak.

Chapter 17 The Poisson distribution

1 (a) 0.175467   (b) 0.175467   (c) 0.006738

  (d) 0.497158   (e) 0.040428   (f) 0.734974

  (g) 0.125110   (h) 0.811936

2 7

Chapter 18 The normal distribution

1 (a) 0.2939   (b) 0.1510   (c) 0.3289

  (d) 0.2417   (e) 0.4931   (f) 0.5

  (g) 0.1562   (h) 0.6293   (i) 0.1587

  (j) 0.5987   (k) 0.04975   (l) 0.3417

2 (a) 0.4772   (b) 0.1464   (c) 0.6826

  (d) 0.1587   (e) 0.5   (f) 0.5398

3 (a) 0.5077   (b) 0.0314   (c) 0.0075

Chapter 19 The binomial distribution

1 (a) 0.273   (b) 0.031   (c) 0.004   (d) 0.492   (e) 0.727   (f) 0.965

2 (a) 0.302   (b) 0.026   (c) 0.107   (d) 0.503   (e) 0.893

3 0.624

Chapter 20 Introduction to hypothesis testing

1 The mean is 4. Various choices of three students can give sample means of between 1 and 9, and so can be very unrepresentative.

2 The following answers are only approximate, from the table:

(a) -1.28 to 1.28 (b) -1.645 to 1.645 (c) -2.575 to 2.575

Chapter 21 z-tests

1 z-test statistic is 1.8974. A one-tailed test. We have evidence at the 5% level but not at the 1% level that the bulbs last longer.

2 z-test statistic is -1.4142. A one-tailed test. There is not enough evidence for the claim at the 5% level (and hence at the 1% level) that the fertiliser is effective.

3 z-test statistic is 1.7330. A two-tailed test. There is not enough evidence at the 5% level (and hence at the 1% level) that there is any difference between the products.

4 z-test statistic is 3.3425. A one-tailed test. We have evidence at both the 5% level and the 1% level that reading the textbook gives better marks.

Chapter 22 t-tests

1 t-test statistic is -2.8868. A one-tailed test. Accept the alternative hypothesis at both the 5% and 1% levels, and so there is strong evidence the pill reduces heart rate.

2 The mean of the differences is 1 and the standard deviation is 2. The t-test statistic is -1.4142. A two-tailed test. Not enough evidence at either the 5% or 1% level to reject the null hypothesis, and so we do not have evidence that eating the cereal bar helps.

3 t-test statistic is -1.897. A one-tailed test. Accept the alternative hypothesis at the 5% level, but not at the 1% level, and so there is some evidence that the fertiliser works.

Chapter 23 χ^2 tests

1 4 degrees of freedom, test statistic is 9.3, critical value is 9.488, so not enough evidence at the 5% level, and the sample is acceptable to be sold.

2 2 degrees of freedom, test statistic is 18.8, critical value is 9.210, so enough evidence at the 1% level that the batch is unacceptable.

3 8 degrees of freedom, test statistic is 154.875, critical value is 20.090,

so strong evidence at the 1% level that there is a correlation between attendance and grade.

4 The same answers are obtained using this formula.

Chapter 24 F-tests

1 The F-test statistic is 4. Using the tables for 10 and 10 (the two sample sizes), the 5% critical value is 2.98 and the 1% critical value is 4.85. We have evidence at the 5% level that the variances are different, but not enough evidence at the 1% level.

2 The F-test statistic is 15.247 (to 3 decimal places). Using the tables for 12 and 15 (the two sample sizes), the 5% critical value is 2.62 and the 1% critical value is 4.01. We have strong evidence at both levels that the variances are different.